THE INVISIBLE WORLD

STUDIES IN INTELLECTUAL HISTORY AND THE HISTORY OF PHILOSOPHY

M. A. Stewart and David Fate Norton, Editors

This is a monograph series whose purpose is to foster improved standards of historical and textual scholarship in the history of philosophy and directly related disciplines. Priority is given to studies that significantly advance our understanding of past thinkers through the careful examination and interpretation of original sources, whether printed or manuscript. Major works and movements in philosophy reflect interests and concerns characteristic of a particular age and upbringing, and seemingly timeless concepts may vary with the changing background of knowledge and belief that different writers assume in their readers. It is the general editors' assumption that a sensitivity to context not only does not detract from the philosophical interest or rigor of a commentary but is actually essential to it. They wish to encourage studies that present a broad view of a subject's contemporary context, and that make an informative use of philosophical, theological, political, scientific, literary, or other collateral materials, as appropriate to the particular case.

Other Books in the Series

Steven M. Nadler, *Arnauld and the Cartesian Philosophy of Ideas*
Catherine Wilson, *Leibniz's Metaphysics: A Historical and Comparative Study*
Thomas M. Lennon, *The Battle of the Gods and Giants: The Legacies of Descartes and Gassendi, 1655–1715*

THE INVISIBLE WORLD

EARLY MODERN PHILOSOPHY AND
THE INVENTION OF THE MICROSCOPE

Catherine Wilson

PRINCETON UNIVERSITY PRESS   PRINCETON, NEW JERSEY

Library of Congress Cataloging-in-Publication Data

Wilson, Catherine, 1951-
    The invisible world : early modern philosophy and the invention of the
microscope / Catherine Wilson.
        p. cm. — (Studies in intellectual history and the history of philosophy)
    Includes bibliographical references and index.
    ISBN 0-691-03418-4
        1. Philosophy and science—Europe—History—17th century.   2. Philosophy,
Modern—17th century.   3. Microscopes—Europe—History—17th century.
4. Europe—Intellectual life—17th century.   I. Title.   II. Series.
B67.W55   1995
113—dc20   94-24556   CIP

This book has been composed in Sabon typeface

Princeton University Press books are printed
on acid-free paper and meet the guidelines
for permanence and durability of the Committee
on Production Guidelines for Book Longevity
of the Council on Library Resources

Printed in the United States of America by Princeton Academic Press

10 9 8 7 6 5 4 3 2

To Mohan

Contents

Preface _____

THE PRESENT BOOK, an analytical essay on the early modern history of science, owes its origins to a study of the preface to Robert Hooke's *Micrographia* undertaken in a seminar on reappraisals of the scientific revolution under the direction of Robert S. Westman at UCLA in the summer of 1983.

Microscopes were introduced into Europe in the first decade of the seventeenth century and were widely manufactured, used, and written about over the following hundred years. Hooke's *Micrographia*, one of the most admired books of the mid—seventeenth century, describes and illustrates his experience with an early form of the instrument. The preface is a study in contrasts: one finds in it an emotional call for sober investigation, a visionary appeal to methodical analysis, a rhetorical plea to abandon words for things, and the claim that instrument-mediated perception, which reveals a "new world" to the eyes, will in time enable us to see our world as it really is.

My own book is a selection of events and moments in the early modern period, with occasional brief glances to either side of the hundred years from 1620 to 1720, and a narrative of reactions to them. I have sought to describe the results of the extension of the empirical horizon to the *subvisibilia* and to offer an interpretation of the emergence of modern science that takes this extension as its central event. I offer an "inverted" picture, in which it is not the luminous and remote objects of celestial mechanics but scrapings, bodily fluids, bits of earth, and fragments of tissue that are located at the juncture of descriptive and speculative natural history, or protoscience, and modern science, and the juncture of occult and scientific mentalities. I have concentrated on some aspects of medicine and of what we now know as biology that have sometimes been regarded as laggards relative to the physical sciences, or even argued out of existence for the period in question.

The first, introductory, chapter discusses some problems of periodization, description, and evaluation associated with the notion of a seventeenth-century scientific revolution. Chapter 2 is concerned with knowledge of the occult, considered both as what is hidden or latent and what is suprarational, and chapter 3 with the evolution and employment of magnifying instruments. Chapters 4 and 5 discuss microscope-based theories of generation and contagion, and chapter 6 addresses the topic of microscopical science and human interest, notably the attempts of philosophers to direct empirical results to the support of metaphysics and theology. In chapter 7, I have turned the results of this experiment in inversion back into philosophy by arguing that the controversy between empiricists and rationalists and the revival of old skeptical and idealistic modes and moves in the late seventeenth and early eighteenth centuries were responses to increasing experience—rewarding and frustrating—with the views opened up by the microscope. The persisting philosophical debates over the merits of instrumentalism, driven by experience with increasingly elaborate instruments for visualizing increasingly remote *subvisibilia*, are a modern echo of this issue.

It is a pleasure to acknowledge the many persons who have assisted at all stages of research and writing. Simon Schaffer, Lorraine Daston, and Brian Vickers were an early source of encouragement and ideas. Philip Sloane made available the University of Notre Dame's collection of old-microscope replicas. David Cox spent an afternoon at his lab at the University of Oregon making glass-bead microscopes with me, and Alexander Rüger passed on useful sources of information. Albert van Helden provided exceptionally helpful assistance on successive drafts, and Christoph Lüthy, Christian Legrand, Edward Ruestow, and Alan Shapiro gave comments and criticisms and generously shared the results of their own researches on early microscopy and the history of optics. Their forthcoming publications are eagerly awaited. I regret that I was not able to make broader use of the richly detailed dissertation of Marian Fournier, currently scheduled for publication, which came to hand only after my own writing was substantially completed. Stephen Gaukroger and Lynn Joy commented beneficially on parts of the manuscript, and the suggestions of an anonymous referee for Princeton University Press led to numerous improvements.

I am indebted as well to the National Endowment for the Humanities Division of Summer Seminars and its Travel to Collections Office, the American Council of Learned Societies, the Alexander von Humboldt-Stiftung, and the Oregon Center for the Humanities, all for financial support. The philosophy faculty at the University of Konstanz and especially Jürgen Mittelstrass provided hospitality between 1984 and 1989. I am grateful to the *Journal of the History of Ideas* for permission to reprint material from volume 49, pp. 85–108. For bibliographical and editorial help, I am indebted to Daniel Tkachyk, and for the preparation of the index, to Heidi Northwood.

THE INVISIBLE WORLD

1

Science and Protoscience

WE MIGHT AS WELL begin with what is not in dispute.

In the early and mid–seventeenth century, Western Europe was the site of theoretical and technological achievements that determined its later economic, military, and cultural history. Scientific societies were founded and journals established; new mathematical techniques, notably analysis and algebraic geometry, were invented and applied to physical problems. Forces and masses were quantified, orbital motion analyzed to determine its components, and free fall brought under the rule of law. The Copernican system won widespread acceptance, and the circulation of the blood was established. Instruments for the observation of very distant or very small objects were put into use, as were newly constructed apparatuses for hydrostatic and barometric experiments. A self-consciousness about the study of nature and its benefits, humanitarian and theological, emerged. New discursive forms, the experimental narrative and the apologia for scientific activity, made their appearance.

So much said, controversy begins. Those who agree that there was an acceleration in the growth of knowledge after the turn of the century are divided on whether it is appropriate to speak of it as a scientific revolution or only as an increase in momentum and coordination. Was there a clean break with the past, or rather a return to the mathematizing of the medieval calculators and the direct observation of the ancients? Was the cultivation of personalities alienated from the literary-humanistic culture of the Renaissance a precondition of these changes or a consequence of them?

Discontinuity in the history of science has been studied at every level, from the "paradigm shifts" of Thomas Kuhn to the "epistemological ruptures" of Alexandre Koyré and Gaston Bachelard, and Michel Foucault's accounts of the emergence of new disciplines and discourses.[1] The old narratives of great men and their sudden flashes of insight have fallen into disrepute; the leap—or stasis—is ascribed to the group. The assumption is that human intelligence, curiosity, and independence remain constant in the species, and are expressed in different ways, with

---

[1] See, representatively, Kuhn, *Structure of Scientific Revolutions*; Bachelard, *La formation de l'esprit sciéntifique*; Koyré, *Galileo Studies*, Foucault, *Order of Things*.

more or less vigor, in all times and places, so that the emergence of science here rather than there, now rather than then, should have nothing to do with the talent of individuals. But if the older historiography of science erred in overestimating the powers of individuals, it did not at any rate make the mistake of imagining the zeitgeist as able to produce exactly the people it needed. Whether we consider the leading figures of the scientific renaissance—Nicolaus Copernicus, Galileo Galilei, Johannes Kepler, Francis Bacon, René Descartes, and Isaac Newton— or those whose names evoke only slightly less recognition—Christiaan Huygens, Antoni van Leeuwenhoek, Robert Hooke—or even the rank and file of the scientifically active, we cannot postulate anything other than a reciprocity between social and material conditions and those who experience them and act on them. An environment favors or simply permits the expression of certain abilities, thereby allowing selected individuals to flourish, whose influence reflects back in turn on their surroundings. Their activities both reinforce those background conditions and alter them, for anyone may possess by accident some characteristics—iconoclasms, unorthodoxies, extremisms—that were not originally favored or selected. Thus, the concentration of capital, the competition for prestige among princes and patrons, and the relevance of medicine, surgery, navigation, and the physics of projectiles and materials to organized warfare have all been recognized as contributing to the organization and pace of the acquisition of theoretical and practical knowledge. The fall of Constantinople, the invention of the printing press, the rise of Protestantism, have all been identified as necessary or determining factors in the rise of science. But such background pressures can determine neither the general nor the particular features of scientific knowledge: contingency and individuality must play complementary and sometimes antagonistic roles.

Although the innovations listed above are frequently thought to define the scientific revolution, which is considered both a break with the past and the beginning of an unbroken phase of knowledge acquisition extending into our time, it is not clear that there is an occurrence here that needs explaining. It is not easy to say exactly what was revolutionary in this alleged revolution, what marked such a radical departure from the interest in nature, its patterns and its deviations, that had always been present. What was different in the conceptual organization, the mentality, of the allegedly unscientific centuries preceding the seventeenth? To what extent is the appearance of discontinuity actually a result of the way in which seventeenth-century natural philosophers conducted themselves in their relations with the physical world, and to what extent is it due to new modes of description, self-understanding, and social interaction? That the "nature" they refer to so frequently was not given to them as such but constituted by them as the object of their endeavors is undeniable. But the forms of social organization and verbal presentation that they invented, and the attitude they struck, were nevertheless directed toward an existing subject that might have been overlooked, ignored, or differently constituted for them. Nor is the idea of the scientific revolution as the beginning of what we have now entirely transparent. We may wonder whether the allegedly new attitudes toward nature and the varieties of institutional and individual activity that they inspired have in fact much bearing on what we consider today to be a scientific attitude, a scientific institution, and scientific practice. Was the protoscience of the seventeenth century an immature form of modern science, or was it the form of something else altogether?

Historians can write the history of science in various ways, emphasizing continuity or discontinuity, innovation or mediation. Any coherent narrative requires the author to identify a signal amidst the noise, and this means that not everything can be included. In earlier phases of the academic discipline, it was more evident how the choice of what to include and what to exclude should be made. The historian attempted to filter out the irrational and the unnecessary—superstitions, oversights, lapses, and repetitions—in order to reach a kernel of positive science: a theoretical statement, an observation, a measurement or procedure that has either retained its value in contemporary science or that was a near approximation to one that has. A maximum of difference between premodern science and that of the seventeenth century was thereby elicited, and a maximum of similarity between seventeenth-century science and our own.

But the plenitude of the data permits other choices and other narratives: both the "otherness" of seventeenth-century science and its continued involvement with its past have been convincingly established in the scholarship of the past thirty years. Primary research has acknowledged the importance of nonepistemic factors, including not only the authority of tradition but the satisfaction of distinct political, theological, and existential interests in the pursuit and acceptance of theories. The lapses and oversights, the supposed nonevents of the history of science, have been found to deliver information in the same way as actual sightings, spottings, detections, and identifications. Repetitions—the whole discourse of physical atomism at midcentury, for example—may tell us as much about what a science is not achieving as about what it is: where it has got stuck. The problems of influence and initiative have proved to be more complex than was initially suspected, either by progressive historians (who saw a steady accumulation of knowledge since the Dark Ages) or by those who favored a punctuated model. Transitions between incommensurable theories can often be shown to have been mediated either by secondary and tertiary writers with a conscious or unconscious interest in reconciliation—this interest being the main reason why they are considered secondary or tertiary—or by primary writers who turn out to display a pragmatic tolerance for inconsistency and multiple approaches. From the point of view of positivist historiography, the result of documenting these relationships is a loss of signal in noise. From another point of view, the noise is part of what history is trying to say.

It is not always inappropriate for the historian of science to begin by looking at contemporary science, reading off the features in virtue of which it is science, and then looking in its history for examples of these features. This procedure parallels that of looking for anticipations and preconditions of current doctrine, and one should not hesitate to employ either, emphasizing continuity in both content and method, especially where these have been overlooked. Suppose, then, that one agrees to adopt the term "protoscience" to refer to seventeenth-century activity without adopting in advance any particular position with regard to its relation to earlier endeavors and to actual science, our science. One may try simply to compare the content of protoscience with what people believed before the seventeenth century and what they believe now. This approach, whatever its earlier value, is nowadays rightly criticized as superficial: it is less interesting to know that people once believed that the moon influenced the growth of crops than it is to know why they believed it or disbelieved it, or came to disbelieve it, or came to believe it again.

Modern science, it is often said, is skeptical, methodical, and quantitative. It is engaged in a constant process of self-revision, which it carries out by designing critical experiments involving physical objects and by tabulating and analyzing mathematically the products of its method. Arguably, these three features begin to congregate in protoscience, establishing its claim to be considered as distant from its past and near to its future. One may even argue that what was said was as important as what was done: it is the forceful and striking introduction of the discourse of skepticism, methodicality, and quantification—whether or not its implications were broadly realized in the practices of seventeenth-century protoscience—that establishes its claim. Although mathematical accuracy and ingenuity were exemplified in the painstaking researches of fourteenth-century medical astrology, skepticism was missing. Skepticism was widespread in sixteenth-century letters, but it was detached from any positive program and constituted in fact the main line of defense of the literary-humanistic culture that Descartes and Marin Mersenne wanted to overthrow. Ramist methodicality, for all its positive pretensions, remained in the realm of language.

The danger nevertheless is that a myth can be created, the myth of the scientific spirit or the scientific mentality, according to which, as in the biological doctrine of preformation, science—our science—was already there in the scientific revolution, invisibly delineated, covered and obscured by adhesions and integuments, but growing and developing as an individual being, to emerge later in its full-blown form. Even if we cannot point to a single figure of the period whose consistent rejection of inherited truth, and whose ability to quantify and to produce potent generalizations describing the nature and behavior of real or theoretical entities, strike us as exemplary from the modern point of view, this only means that the scientific spirit was restrained, or frustrated, or subject to interference in various ways.

This is a claim that has not infrequently been made—science is *improbable*.[2] The problem lies in determining exactly who or what was restrained. For science does not seem to be the undertaking of any particular agent, who might experience it as difficult, who might have taken it up against the odds. It is rather the sum of undertakings of a number of agents whose cooperative activities seem to render it likely and facilitate its development. Admittedly, the activities of any single agent are fraught with difficulties, conceptual, technical, and social: Robert Hooke's career provides a fine example of how a personality can impede a career, and of what can be accomplished anyway. Part of what is meant in saying that science is unlikely and arduous is clear and indisputable: not every culture, left to itself, will become, like ours, a culture of science. But there is more to the claim that science is difficult and unlikely than this. Those who view the history of science as a series of conceptual innovations practiced by people standing outside the mainstream of their time are especially disposed to register a certain impatience with science as it was actually practiced, to feel that it was being impeded in some way, or held back, by the rank and file of not particularly gifted ordinary practitioners, or by the persistence of outworn conceptual frameworks or other "obstacles."

One such alleged obstacle is a perceived conflict between two modes of explanation, one directed toward exact mathematical description, the other toward the explanation of particular phenomena. According to Richard S. Westfall, "the full fruition of the scientific revolution required a resolution of the tension between the two dominant trends."[3] By resolution he means the successful importation of quantitative

[2] For example, Gillispie states early in his study of the scientific revolution that science is "a most arduous and unlikely undertaking." *Edge of Objectivity*, p. 9.

[3] "Two major themes," he argues, "dominated the scientific revolution of the seventeenth century—the Platonic-Pythagorean tradition, which looked on nature in geometric terms, convinced that the cosmos was constructed according to the principles of mathematical order, and the mechanical philosophy, which conceived of nature as a huge machine and sought to explain the hidden mechanisms behind phenomena." Westfall, *Construction of Modern Science*, p. 1.

methods into such qualitative, descriptive fields as chemistry and
what we now call biology—a field drawing on a number of older
branches of inquiry: anatomy, natural history, medicine, and natural
philosophy or "physics" in the old sense. In a similar vein, Charles
Coulton Gillispie argues that it is mathematical physics, in the modern sense of the term, that had always had but finally was recognized
as having pride of place: "Physics has been the cutting edge of science
since Galileo, and its mathematicization in dynamics was, therefore,
the crucial act in the scientific revolution."[4] Once again we have an
ontology of acts, acts that are neither those of a particular person nor
those of any identifiable collective. More abstract still is the ontology
of Koyré, for whom thoughts themselves have a formative power:
"[I]t is thought, pure unadulterated thought, and not experience or
sense-perception, as until then, that gives the basis for the 'new science' of Galileo."[5] And again: "[O]bservation and experience, in the
sense of brute, common sense experience, did not play a major role—
or if it did, it was a negative one, the role of obstacle—in the foundations of modern science."[6]

Obstacles, impediments, tensions!—but are not these troubles of
early modern science really in fact the projected troubles of the historian who finds it difficult to put the story together coherently? Once it
has been decided that the real contributions of the seventeenth century
were its image of nature as a machine, or (a different matter entirely) its
image of nature as something related to a machine as a clock face is
related to the clock mechanism, and its desire to formulate mathematical laws governing mechanical and dynamic interactions, then protoscience, especially in its chemical, physiological, and medical branches,
will inevitably be seen as a failure to get on with it. It will be tempting
to explain this failure by reference to imaginary internal tensions, tensions experienced by no person, and imaginary impediments that no
demon has placed in the way. The resulting account is dramatic, like all
contests of force and clashes of will, but is it correct?

For all the criticism such interpretations have suffered in recent years,
they retain something of their grip. They are associated, moreover, with
a certain picture of seventeenth-century scientific activity, one that is
naggingly familiar and not entirely unsalvageable. Seventeenth-century
science, on this scheme, broke with ancient habits of thought and regu-
lative assumptions; its underpinnings were a new philosophy. It abandoned the qualitative universe of Aristotelian substances, in which there

---

[4] Gillispie, *Edge of Objectivity*, p. 54.
[5] Koyré, *Metaphysics and Measurement*, p. 13.
[6] Ibid., p. 18.

exist natural places, purposes, and hierarchies, in which harmony and above all appropriateness and suitability are reasons why something should be or happen. Nature is now revealed, on this account, as something elusive and sufficient unto itself, not, in Gillispie's choice phrase, "an enlargement of common sense arrangements, not an extension of consciousness and human purposes."[7] In Aristotle's physics, each kind of flesh, each mineral, each drug has a particular form and particular qualities; the moderns are said to recognize only homogeneous corpuscles with figure, size, and motion, and to explain all qualities and powers as the effects of fit and interaction between these particles. The objectivity of science, on Gillispie's account, is supported by its dissociation from human interests; heliocentrism, mechanism, and the reduction of qualities are the theoretical developments on which this dissociation is itself based.

It is not hard to call to mind in connection with this hypothesis, or with the hypothesis that mechanism and mathematics were the driving forces of early modern science, certain works that seem to establish the tone of the period: Galileo's *Discourse concerning Two New Sciences* (1620); Descartes's *Principles of Philosophy* (1644); Robert Boyle's *Excellency and Grounds of the Mechanical Hypothesis* (1666); Newton's *Mathematical Principles of Natural Philosophy* (1687). But even within these specially selected works, and in the minds of their authors, there are pieces that do not fit the prototype. Descartes may enunciate the reduction of the world to extension and motion, but the corpuscles that figure in his protoscience are still qualitatively distinguished into the earthy, the airy, and the fiery in a way hard to reconcile with his claim that matter is only extension. Boyle's pronouncements in favor of mechanical philosophy similarly move on a different level from his experimental work, and Newton's belief that gravity is a manifestation of the omnipresence of God and his interest in vital principles show that anthropocentrism and existential interests can happily coexist with successful mathematical abstraction. Nor is there apparent in these works any sign of tension between reductive causal explanation and mathematical systematization. From the agent's perspective, it seems rather that it all fits together as it should. Descartes fails to provide any rigorous, nonarbitrary deduction of the micromechanical models he proposes, but this is not a problem of employing accurate numerical values in explanations of particular phenomena. To locate such tension, we must imagine the results of our trying to map Cartesian mechanism or Newtonian dynamics onto medicine or chemistry; but in fact Cartesian iatromechanics and Newtonian medicine had their confident propo-

---

[7] Gillispie, *Edge of Objectivity*, p. 13.

nents as well as their opponents, who detected in these theories not internal tensions but absurdity.

In summary, the focus on physical science and the mathematization of the laws of motion constitutes a selection of the historical data that effectively prejudges the question of the relation of protoscience to science by separating signal from noise at the outset. This account, based as it is on the sameness of physics and the difference of natural history, medicine, and chemistry, discourages us from examining contemporary appraisals of the limitations of the mechanical philosophy and the wide vistas opened by a nonrigorous, unsystematic empiricism. It conveys the impression that what was responsible for the general air of confidence and progress, the proliferation of societies and publications, the manufacture and sale of instruments, and the formulation of certain patterns of accommodation between religion and science were atomism, mechanism, and mathematization, and that philosophy—an a priori analytical discipline telling us how things must be—provided their foundation.

The ubiquity and abundance of philosophy, methodology, and systems in the seventeenth century impressed the first generation of intellectualist historians, Koyré and Edwin A. Burtt. Reacting against earlier presentations that emphasized the accumulation of facts, the rejection of superstition, and the independence of science from religion and speculative philosophy, they plotted out a metaphysical groundwork for the scientific worldview. Reciprocally, the recent history of philosophy has left behind the exercise of mapping entailment relations a priori and now frequently emphasizes the dependence of philosophical theory on empirical results. We have learned from studying the relation between Descartes's and George Berkeley's theories of vision and their metaphysics, from comparing Gottfried Wilhelm Leibniz's concern with the conservation of forces with his theory of monads, to see philosophers' work as having both a conceptual and an experimental dimension. A justified dissatisfaction with the anachronistic term "scientist" has produced as a replacement "natural philosopher," thereby making unavoidable the implication that scientific work depended on acquaintance with and conditioning by philosophical views, and vice versa. These developments are commendable, insofar as they avoid reading history as though it respected disciplinary boundaries actually imposed much later. And the interaction between metaphysics and observation was as pronounced outside the physical sciences as within them. But there is a danger that, in acknowledging exchanges between natural science and philosophy, one ignores their hostilities and differences. Here it is not a case of imaginary tensions, but of documentable dissatisfactions.

The question of the role of philosophy and of foundational theories is complicated by the use of the term "philosophy" in two distinct senses, a pejorative one and a neutral one. William Whiston in his survey of the Oxford curriculum in 1717 states: "No uncertain systems of Philosophy to be recommended; but Mathematicks and Experiments to be prefer'd."[8] The contrast between philosophy on the one hand and mathematics and experiments on the other should be taken to heart. Mathematics, in 1717 and earlier, was recommended as a subject of study not for its intrinsic certainty, or because it was thought to provide a scaffolding for a world reduced to pure relations of quantity, but for its practical applications. John Wilkins's *Mathematical Magic* of 1648, a compendium like Gaspar Schott's famous *Technica curiosa* of 1664, is really a treatise on calculating techniques and machines. Certainly philosophies had their place as prolegomena. The Oxford curriculum of 1707 names the corpuscular philosophy, to be approached via Descartes, Boyle, Jacques Rohault, and Jean Le Clerc. But there is no indication that philosophical theories concerning matter, qualities, or the laws of nature were ever intended to be studied as the foundations of physics. The student simply studied Descartes's *Principles of Philosophy*, along with optics, conic sections, and curves learned through Kepler, Descartes, Huygens, William Molyneux, Isaac Barrow, and Newton, and chemistry through Boyle, Nicolas Lemery, and others. The student was also supposed to familiarize himself with the micrographical studies of Hooke and Leeuwenhoek on minerals, of Nehemiah Grew and other contributors to the English *Philosophical Transactions* and the German *Miscellanaea curiosa* on the microscopical and macroscopical features of plants, and various anatomical papers on animals.[9] Further, the curriculum prescribed astronomy and "mechanical philosophy"—statics, hydrostatics, collision mechanics ("percussion"), gravitation, and tidal phenomena. From these lists it is apparent that, if protoscience did have metaphysical foundations, they were not taught and learned as such. The term "mechanical philosophy" did not typically refer then, as it does in modern philosophical usage, to corpuscularianism, universal mechanism, and the associated theory of primary, secondary, and tertiary properties, but to studies that could be undertaken of, or with the help of, machines.

Newton's *Opticks*, Thomas Burnet's *Sacred Theory of the Earth*, the physics and anatomy of Rohault, Thomas Bartholin, and Caspar Bartholin, the geography and the astronomy of Wells, Whiston, Gregory, and Keill are all referred to in the curriculum as "Philosophy," which

[8] Gunther, *Early Science in Oxford* 3:338.
[9] Ibid., p. 387. William Harvey and Giovanni Borelli are mentioned.

comprises Corpuscular Philosophy, Experimental Philosophy, Anatomy and Philosophy, and Mechanical Philosophy. This was philosophy in the neutral sense. These were systematic and even speculative works, but they were not philosophy of the form thought to be useless and uncertain. "Natural philosophy" differed both in content and in form from metaphysics—from philosophy as traditionally understood. It did not hold itself responsible for contributing to knowledge about God, the soul and its destiny, substance, inherence, modes, being, time, infinity, and eternity. Descartes's textual engagement with these topics is simply an attempt to accommodate previous expectations of philosophy while subverting them; in his *Meditations*, he draws a theological and metaphysical picture frame around his physics. The seriously devout, such as Boyle and Nicolas Malebranche, do something not much different and reserve their moral and theological concerns for other works. Physico-theology, which might seem to constitute an ideal blend of empirical observation with religious hope, was also a frame, available to the indifferent as well as the devout. Even Leibniz, who cared deeply about the traditional concerns of metaphysics, could continue to talk about them only by producing two worlds: an unperceived metaphysical world outside time and space, and a phenomenal world where mechanics held sway.

In place of the intellectualist view that atomism and universal mechanism, established and sustained a priori and with the help of a theology of divine command, were the foundations of protoscience in its more satisfactory manifestations, I would argue that these doctrines were construed in various ways, none of which is precisely foundational. They could be regarded as philosophy in the pejorative sense, as fanciful commitments to a simplified ontology that transcended experience, as opponents of Descartes such as Joseph Glanvill did. Alternatively, they could be construed neutrally as conceptual replacements for philosophy in the old sense to which one need not adhere dogmatically: this seems to have been the view of John Locke. Or else—and this construction will receive full attention in due course—atomism and mechanism could be construed as outside the realm of philosophy altogether, as given in experience.

Both positivists, who argue that science became science when it set philosophy aside, and intellectualists, who believe that philosophical reconceptions were essential to its development, agree on a certain doubtful assumption: that truth may emerge out of error but cannot emerge out of confusion. For the positivist, the application of canons of evidence and scientific protocols determines the forward movement of science. For the intellectualist, novel theories are recommended by their perspicuity, elegance, and economy. On either account, those who did not have a clear picture of their own assumptions, hypotheses, and convictions, and from whose researches protocols and theories of method cannot easily be extracted, were practitioners of Baconian science, which is characterized by haphazard collecting and showy but vaguely conceived "experimentation." And the sciences that were practiced in this way—sciences other than astronomy, statics, and dynamics—are held to have lagged behind, not to have been constitutive of, the scientific revolution.

There came a time when the Baconian approach to certain subjects was obsolete and amateurish. But the perception of a lag in seventeenth-century chemistry and biology, to use the anachronistic but necessary term, resides perhaps in the eye of the beholder. The new science curriculum within the universities recognized these areas, and the philosophical societies outside the universities did so to an even greater extent. Yet it is not uncommon for historians to deplore the state of biology in the late seventeenth century, lamenting that it was either too much like physics or, alternatively, too unlike physics to make satisfactory progress.[10] The failure of biology to become sufficiently mechanical, or sufficiently vitalistic, and the failure of chemistry to become quantitative are imaginary failures. Francis Glisson's *Tractatus de natura substantiae energetica* of 1672 might seem to determine, together with Giovanni Borelli's *De motu animalium* of 1681, a field of possible future biologies ranging from the "vitalistic" to the "mechanistic." Yet these works stand out in the mass of contemporary literature as representing mere possibilities: as reveries. In the latter, the body is presented as a mechanico-hydraulic system of canals, sieves, filters, and valves on the one hand, and levers, pistons, and centers of gravity on the other.[11] For all its attention to the hydrostatics of fish and the physics of sound, *De motu animalium* is, by later standards, a fantastic work of the analogical imagination. To see in it the route regrettably not taken until later is to miscategorize it; it is entirely less modern than conventional textbooks of anatomy of the period.

Such fancies, like the "living machine" described in Descartes's posthumously published *De homine* of 1662, can exert an influence. Marcello Malpighi wrote a now lost work in the late 1650s consisting of a dialogue between a Galenist and a mechanist, and claimed in his inaugural lecture at Messina in 1662, apparently arousing little opposition,

[10] Compare here Gillispie's account in *Edge of Objectivity* with the argument of Roger, *Les sciences de la vie*.

[11] See Belloni, "Schemi et modelli della macchina vivente."

that the study of anatomy had been put onto a firmer footing by mechanism and enriched by the discoveries of Descartes.[12] But it would be wrong to conclude that the mechanical theory of the organism was constructed out of whole cloth by Descartes, developed and refined by Borelli, delivered to his pupil Malpighi, and regrettably abandoned. Such a story would ignore the inbred and acquired hostility to speculation and theory, the assertion of a right to ignore foundational questions, that a fuller account of the three men's intellectual relations makes clear.

Borelli had been impressed by Galileo's belief in an underlying mathematical order to nature and by the "free and Democritean" philosophy of Pierre Gassendi.[13] He came to the conclusion that vital phenomena are performed not by a soul but materially and mechanically. He sought through the practice of anatomy to confirm this view, though he was incapable of making his own dissections and prevented by his poor eyesight from effective microscopical examination of the animal anatomy. By 1663, having always found Gassendi more congenial than Descartes, Borelli was writing disparagingly about the latter. "A new book by Descartes has appeared, called *De homine*," he tells Malpighi, "but he is so nauseating to me that I shall easily dispense with looking at it."[14] Trying to discourage Malpighi's admiration for the Cartesian philosophy, which he derides as full of "extravagances," he warns him that "your respect for Descartes must not keep you from speaking quite freely, because he is not so highly esteemed as you imagine; on the contrary, his credit and reputation are continually on the ebb."[15]

By 1665 Malpighi's enthusiasm too had cooled. He not only denied being influenced, he rejected any role for Cartesianism, by which he meant speculation and theorizing in anatomy. "I have come to know this structure of the kidneys," he says in the proemium to *De renibus* (1666), "not by any means through the use of books, but solely through the patient, long-continued and varied use of the microscope, and . . . what else has been deduced from this, I have, as is my wont, contrived by my own slow intelligence and remiss mind."[16] This remark cannot be taken entirely at face value; the rhetorical opposition between seeing for oneself and reading in a book, between the patient observation of detail and the hasty exercise of genius, are conventional and exaggerated. Descartes's great systematizing works, the *Principles of Philosophy* and the *Treatise of Man*, with their statements of universal mechanism, were

[12] Malpighi, *Opera posthuma* 1:25, in Adelmann, *Marcello Malpighi* 1:211.
[13] Adelmann, *Marcello Malpighi* 1:151.
[14] Letter to Malpighi, 30 March 1663, in Adelmann, *Marcello Malpighi* 1:217.
[15] Letter to Malpighi, 16 May 1664, in Adelmann, *Marcello Malpighi* 1:238.
[16] Malpighi, *Opera posthuma* 2:278, in Adelmann, *Marcello Malpighi* 2:820.

not wholly cut off from observation and practice in dissection. Malpighi exaggerates his own indifference to books and theories; he did step in when attacked to defend mechanism and to make his own orientation more explicit. Yet his antiphilosophical stance is typical. Modern criticism has discovered the "natural philosophers" in those previously thought to be transfixed in their armchairs, but it has wrongly assumed that science was as interested in philosophy as philosophy was in science. The avoidance of theory should, however, give no comfort to the positivist who believes that modern science began with the development of specific procedures for separating truth from falsity. Certainly we find, above all in Bacon, a statement of the need to develop such procedures, and even a master procedure for discovering all forms. But in practice, Baconian method became Baconian empiricism, the paradigm of unsystematic, ungrounded science. Neither procedures nor metaphysics seem to play a determinate, uncontested role in the emergence of a new science.

And why should science not have emerged from a haphazard and chaotic empiricism, from experience with making, mixing, and measuring, simply under pressure, as Aristotle says, from the truth?

One reason—not perhaps the most fundamental—why this suggestion tends to produce a sigh of despair is that a revolution in foundations or procedures seems necessary to explain the timing of what has come to be called the scientific revolution. Observation of nature, instrument making, and even interventionist experimentation had been practiced earlier by the ancient naturalists and astronomers, in the medieval period, and in the early Renaissance, by alchemists, doctors, druggists, and compilers of books of household "secrets," without resulting in the accelerated developments of our period or in the consciousness of a scientific movement. Meanwhile, the medieval curriculum was fixed in its trivium and quadrivium, with the study of logic, rhetoric, and grammar, astronomy, arithmetic, music, and geometry. It is supposed that, without the proper philosophy to underwrite the effort to reject this curriculum, and a new conception of method to substitute for scholastic logico-linguistic analysis and dialectic, there could neither be self-consciousness of science as an enterprise nor progress. But one should not forget that this curriculum was already, before the appearance of the great philosophical reformers and methodologists, being eroded simply through the recovery and distribution of ancient texts on natural science in the fifteenth and sixteenth centuries. Ulisse Aldrovandi at the University of Bologna stretched the boundaries of the traditional curriculum by lecturing and publishing on medicinal plants, fossils, and animals in 1561; Girolamo Cardano, Julius Caesar Scaliger, and Hieronymus Fabricius were all active in a period that had nothing of modern philosophy behind it.

A second reason for supposing that theory and methodology determined the brisk pace of development in early science is that the philosophy of science has become steadily less friendly to the view that truth will out even in the absence of adequate foundations or procedures. The history of science was formerly presented as an incremental process, in which facts were unearthed and stored in the treasure-house of human achievement, from which the corresponding errors were carted away. The emphasis now is on the dependent and relative status of facts and observations themselves, which are said to have meaning only within a wider theory, and to achieve their status only in relation to other assumptions and background conditions. From this perspective, progress requires the replacement of one comprehensive view—which necessarily implies an ontology of beings and actions—with another view that is equally or more comprehensive. Science is given its dynamism, and its very substance, by a penumbra of speculations and convictions that go beyond its official pronouncements and bring it to life. Even those who did not profess an interest in Cartesianism or Boylean corpuscularianism, it might be argued, showed by their very behavior, by the problems they set for themselves, that the mechanical worldview was the frame and the background for their researches.

A third consideration, perhaps the most substantial, that has elevated mechanism and mathematics to the status of major explanatory categories for the rise of science is the belief that the seventeenth century created for itself an ideal of objectivity. In doing so, the anthropomorphic and vitalistic elements of Aristotelianism—its four causes, its teleology, its powers and virtues conceived on analogy with human dispositions—were expunged. A desire arises, it is said, to comprehend the world as it is in itself, not as we see and feel it to be, and this, it is said, is what science is really about. Galileo's proposal that sensory qualities are the result of interactions between the human mind and small particles is followed by Descartes's clear distinction between the level of physical existence, where we are concerned with bodies of various shapes, sizes, and motions, and the presentations to consciousness conjured up by mind-body interactions.[17] Robert Boyle's destructive analysis of the "vulgarly received" notion of nature similarly appeals to "one catholick and universal matter common to all bodies . . . a substance extended,

---

[17] "Light, colour, smell, taste, sound, hot and cold," as well as the other tactile qualities, are "simply various dispositions in those objects which make them able to set up various kinds of motion in our nerves which are then required to produce all the various sensations in our soul." Descartes, *Principles of Philosophy* 4:199, in *Philosophical Writings* 1:285.

divisible, and impenetrable."[18] His attack is directed against the notions of form, power, and virtue, the unifying charge being the subjectivity and unintelligibility of these notions. To think of nature as a power, as a set of inherent forces, is to deify it, for "according to our hypothesis, inanimate bodies can have neither appetites nor hatreds, nor designs, which are all of them affections, not of brute matter but of intelligent beings."[19] Struck by these passages, interpreters of the scientific revolution have argued that the creation of a notion of objectivity was essentially tied to scientific reductionism and the banishment of phenomenological properties from the realm of what exists apart from human observers. Burtt, in a memorable passage describing what he calls "simply an incalculable change in the viewpoint of the world," states:

> The scholastic scientist looked out upon the world of nature and it appeared to him a quite sociable and human world. It was finite in extent. It was made to serve his needs. It was clearly and fully intelligible, being immediately present to the rational powers of his mind; it was composed fundamentally of, and was intelligible through, those qualities which were most vivid and intense in his own immediate experience—color, sound, beauty, joy, heat, cold, fragrance, and its plasticity to purpose and ideal. Now the world is an infinite and monotonous mathematical machine. Not only is his higher place in a cosmic teleology lost, but all those things which were the very substance of the physical world to the scholastics—the things that made it alive and lovely and spiritual—are lumped together and crowded into the small fluctuating and temporary positions of extension which we call the human nervous and circulatory systems.[20]

The old controversy between Aristotle and the atomists is replayed, this time with a different outcome. Democritus had said that "by convention are sweet and bitter, by convention is colour; in truth are atoms and the void."[21] This view had been challenged by Aristotle as failing to do justice to experience: qualities could not come up from nowhere, and philosophers should be wary of the tendency to deny what we see and feel. Yet here it is again, now triumphantly established. The world is not there to delight us; the scientist is one who unmasks the delusions of self-indulgent human consciousness and replaces them with the hard facts: nature in itself is morally and aesthetically neutral, neither benevolent nor cruel, neither beautiful nor ugly. In place of a sympathetic cosmos, whose members are bound together by analogies, harmonies, and sympathies and kept distinct by metaphysical individuality and an-

[18] Boyle, *The Origin of Forms and Qualities*, in *Works* (1772 ed.), 3:15.
[19] Boyle, *Free Inquiry*, in *Works* (1772 ed.), 5:222.
[20] Burtt, *Metaphysical Foundations of Modern Science*, p. 116.
[21] Kirk and Raven, *Presocratic Philosophers*, p. 422.

tipathy, we have only one kind of matter forming one pattern, and, in place of love and strife, little structures and machines producing all the illusions of subjectivity. As Walter Charleton put it in midcentury, "in every Curious and Insensible Attraction of one bodie by another, Nature makes use of certain slender Hooks, Lines, Chains, or the like intercedent Instruments, continued from the Attrahent to the Attracted, and likewise . . . in every Secret Repulsion or Sejunction, she useth certain small Goads, Poles, Levers or the like protruding Instruments, continued from the Repellent to the Repulsed bodie."[22] There is no action at a distance: sentiment cannot be physically efficacious.

This disenchantment of the world picture is supposed to be accompanied by a sort of mental maturation. If nature is simple and lawful, it is possible for those able and willing to accept these abstractions to apply themselves to learning the underlying code, and so to derive a higher form of intellectual satisfaction than the naive sensory and imaginative satisfaction that comes from a belief in natural purpose, human centricity, and real qualities. If there is more than a trace of nostalgia in Burtt's description of the more exacting, less friendly world of things in themselves, grand and forbidding by contrast with the world in which we are or were at home, there is nevertheless a moral lesson in this about the burdens of privilege. The coldness and aloofness of the scientist are the price that has to be paid for exceptional vision. One of Descartes's favorite points is that the senses furnish only a servile sort of knowledge, one that allows us to survive and that is possessed by every infant and every peasant, not the kind of knowledge that both glorifies the knower and produces cultural progress.[23] Through philosophy, we must get behind appearances to their causes and foundations. "In our childhood, our soul or our thought was so strongly impressed by the body that it knew nothing distinctly, even if it perceived some things clearly enough; and because it did not refrain from making judgements on what presented itself our memories are filled with numerous prejudices, from which we hardly ever try to deliver ourselves."[24] And this disparaging view of ordinary experience is often associated in the literature with attacks on Aristotle and his followers—"those Sons of Sense," as Henry Power calls them.

I am concerned with how the alternating validation and rejection of both ordinary and lens-assisted sense perception is to be understood: Descartes's own oeuvre is awash with countercurrents in this respect. It

[22] Charleton, *Physiologia Epicuro-Gassendo-Charltoniana*, p. 344.
[23] Cf. Gillispie on "aristocratic" and "democratic" knowledge, in *Edge of Objectivity*, p. 75.
[24] Descartes, *Principles of Philosophy* 1:47, in *Philosophical Writings* 1:208.

suffices to note here, where my concern is with the historiography of science, how this language is interpreted to support the picture of science as a difficult and acquired taste, and theoretical physics as the most difficult and acquired taste of all, in its purity, its abstractness, the decisiveness and beauty of its results, and its irrelevance to the ordinary problems of survival. It is not humane, on this view, and does not need to be, because it is true. The Baconian sciences lend themselves readily to popularization, they attract amateurs, and they exist thereby in a degraded state. Although even scientists who are not Baconians like to represent themselves as motivated by humanitarian ends, this descent to the enclosed domestic world of human purpose is only a pretended concession. Newton, "with his prism and silent face," as Wordsworth put it, is for Gillispie an emblematic figure. Science is self-sufficient, confident about the value of what it has. It has no need to plead for itself, to become benevolent and communicative.

What counts for the intellectual interpreter of the scientific revolution is the relinquishment of the superficial human perspective, a relinquishment that may not really have occurred even when we seem to be getting underneath surfaces, even when the point of view seems to be other than the ordinary human one. Andreas Vesalius's achievement in his *Fabric of the Human Body* (1643) was not, Gillispie argues, a significant act of the scientific revolution. These figures of skeletons and flayed men set against a Tuscan landscape belong to the same category as Leonardo's studies. There is nothing of conceptual innovation in them; this is simply a repackaging of Galenic doctrine. Neither Vesalius nor any biologist before Darwin produced ideas, he says, "which changed man's conception of the world or even of himself."[25] But in fact there is conceptual innovation in the broad display to the eyes of the inner anatomy of the human body. Anatomical illustration gratified a hunger for visual experience, but this did not prevent it from contributing to the development of objectivity. The claim that nothing of general significance happened in biology before the Darwinian "assimilation of biology to the objective posture of physics" is one I believe is false both in its estimation of biology and in its assumptions about objectivity.

Unlike the labors of the mathematician, Baconian science is subject to absurd contingencies. Bacon realized this himself, noting that it was widely believed that the dignity of the mind is impaired by "long and close intercourse with experiments and particulars, subject to sense and bound in matter; especially as they are laborious to search, ignoble to meditate, harsh to deliver, illiberal to practise, infinite in number, and

---

[25] Gillispie, *Edge of Objectivity*, pp. 57–58.

minute in subtlety."[26] The rambling accounts of Robert Boyle remind the reader that broken glassware, spills and stains, cluttered tabletops, lost notes, failed experiments, explosions, and irreproducible results are features of scientific experiences. And the results! Baconian science rapidly became a parody of itself: Restoration satirists were allegedly able to lift whole passages out of the *Philosophical Transactions* and insert them into standard romantic plots.[27] It was as funny to its contemporaries as it was to later browsers in the *Transactions*, and the dignity and seriousness of the aristocratic members of the early Royal Society were part of the humor.

Consider again the suggestion that the objectivity of science is a consequence of the perspective adopted by the mathematizing atomist who escapes both the inconsequentiality and discursiveness of the superficial performances of idle virtuosi and the illusions of subjectivity. And note first that the reductionist accounts of human perception and the elimination of sympathies and antipathies and other evident projections of the human psyche into nature were not the necessary accompaniments of atomism or mechanism. For Descartes, Galileo, and Boyle, mechanism and corpuscularianism were paired with sensory reductionism, the denial of real qualities. But Gassendi, Charleton, and for a time Thomas Hobbes, equally celebrated as moderns in their day, believed in visual species that they construed materially as icons, films, or *idola*, "decortications or sloughs," which, emitted from the object, flew through the air like snakeskins carrying "an exact resemblance of all Lineaments and colours," as Charleton described them, entered into the eye and mind, and so made us see.[28] The doctrine that qualities are simply the effects of corpuscular arrangement was an additional hypothesis, in which the knowledge that the eye was somewhat like an optical device, a camera obscura that projected a picture on the back of the eye, had to be forgotten. The new view was that the eye was responding to a pattern, as Descartes knew, of pressures: "the sensation of light is the force of the movements taking place in the region of the brain where the optic nerve-fibers originate, and what makes [the soul] have the sensation of color is the manner of these movements."[29]

To what extent did the strong reductionism of Descartes imply an attack on a science based on sense perception? Descartes says at the beginning of the *Dioptrique*, "The entire conduct of our life depends upon our senses, amongst which that of sight is the most noble and

[26] Bacon, *New Organon* 1:lxxxiii, in *Works* 8:115–16.
[27] See Nicolson, *Science and Imagination*, pp. 188–89.
[28] Charleton, *Physiologia Epicuro-Gassendo-Charltoniana*, p. 141.
[29] Descartes, *Dioptrique*, in *Philosophical Writings* 1:167.

universal, so that there is no doubt that the inventions which serve to improve its power are the most useful which can exist." These words, which nonironically recall Aristotle, carry no hint of the later implication that seeing is essentially *servile;* they reveal no ambition to remove oneself from the sensory world, but rather an ambition to bring more within the range of direct perception through the use of optical instruments, which, "carrying our vision farther than the imagination of our ancestors was accustomed to range, seem to have opened up the way to a knowledge of nature greater and more perfect than they ever had."[30] It has recently been argued that the celebrated Cartesian method was not intended to produce a deductive system of the world, but to show how mathematical results could be transformed into practical devices for progress in achieving knowledge.[31] The method was intended to link optical theory to artisanship in the construction of lenses, the construction of lenses to knowledge of the microanatomy of the body, and the microanatomy of the body to medicine. The actual Cartesian science that came of Descartes's efforts was not desired but an unfortunate consequence of his inability to carry through with the first step. The grandiosity of the *Principles of Philosophy* is, on this interpretation, due to its character as a substitute for an experimental science that failed, not because of imaginary obstacles or tensions, but because of the resistance of materials and human weakness. The mechanical models hypothesized for every natural phenomenon, then, became science by default; the distance imposed by the need to conceptualize what could not be seen was not something at which Descartes had ever been aiming.

How firm is the connection between the asceticism that Burtt and Gillispie detect in the stance of objectivity and the banishment of qualities and relations, on the one hand, and the development of early modern science on the other? Besides Galileo's account of warmth and tickling as mind-dependent phenomena in the *Assayer*, other instances come to mind: the mysterious phenomena of love at first sight and instant aversion are explained by the atomist Charleton as the result of invisible streams of particles emitted by one being and reacting on another.[32] There is no sudden impoverishment here, no real ontological strictness where events are concerned, no restriction of what can happen. The corpuscularian philosophy allowed for every sort of action, so long as one was willing to imagine it mediated by mechanisms involving invisible effluvia. Charleton also gives the solution of the basilisk and discusses celestial influence on sublunary bodies; Glanvill in the *Vanity*

[30] Descartes, *Philosophical Writings* 6:81.
[31] Legrand, "De l'invention des lunettes," pp. 15ff.
[32] Charleton, *Physiologia Epicuro-Gassendo-Charltoniana*, pp. 344ff.

of Dogmatizing portrays the dogmatist as the one who rigidly excludes action at a distance. Telepathic communication and "sympathetical limbs" are as much the phenomena of empirical science as magnetism is.[33]

And was the medieval world, given its doctrine of real qualities, distinct substances, and visual species, "those little images flitting through the air," as Descartes called them in the *Dioptrique*, really warm and friendly, alive and lovely? Did the natural world really present itself before the scientific revolution as inviting a pas de deux with the human mind and afterward as a victim stretched on the rack? Europe in the century before the appearance of plague was, one might infer from its art, warm, happy, and certainly colorful. The orderliness of the cosmos in the medieval tradition, the rankings of angelology and the hierarchical system of enclosed spheres, might be supposed to have conveyed a certain snug security. But medieval philosophy was ambivalent toward the natural world and the value of the sense of vision. *Curiositas* was one of the sins most repellent to Augustine; the profound ocular aestheticism of the *Confessions* is the other face of the fear that visuality and spirituality are as mutually exclusive as gratified lust and salvation. Representatives of Christianity, a religion of self-examination stressing not the obedience to external law but the purification of motive, argued that this task was sufficiently absorbing to leave no energy available for external applications of the questioning faculty, a view that persisted until physico-theology and popular science converted the knowledge of nature in an idle culture into a moral pastime far less dubious than the reading of novels. The light of nature was dim, the light of grace radiant. The famous lines from Romans 1:20 to the effect that the invisible things of God are seen in visible things, the things that are made, would eventually be turned toward an apology for scientific curiosity. But it was not a text originally intended as a slogan for research: the point was rather that the things that are made are to be looked beyond; they are symbols and reminders, not the termini of contemplation. Not only was the theme of the essential wickedness of nature important in establishing the logic of the doctrine of original sin, but the theme of its inefficacy was important for eradicating remnants of folk religion and nature worship.[34]

The sensuousness that Burtt ascribes to the Christian culture of the medievals has therefore little to do with its philosophical apparatus. When the scholastic interpreters of Aristotle were criticized, their oppo-

---

[33] Glanvill, *Vanity of Dogmatizing*; see his story of the man who insisted on having his apparently healthy but bewitched arm amputated, in *Scepsis scientifica*, pp. 206–7.

[34] Thomas, *Religion and the Decline of Magic*, p. 256.

nents never mentioned their affective attachment to things of the world, but rather their dullness, their dryness, their vocabulary unconnected to experience, their abstraction, their celibacy. Bacon makes heavy weather of the last especially: the scholastics can talk, he says, but not generate; they are "fruitful of controversies but barren of works."[35] "Shut up in their monasteries," they are unable to make contact with any extralinguistic subject matter. It might nevertheless be said that in order to give itself a face acceptable to theology, science had to become desensualized, and that this was a task carried out by Descartes's chief interpreters. Malebranche's Augustinian tendencies—subject to considerable interference, it must be said, from his *curiositas*—are reflected in his general dismissal of the illusory world of sense, which contrasts with the relatively featureless, simple, regular, mechanical world of scientific reality. It is not the physical world that gratifies the eye and ear and tongue with colors, sounds, and tastes; it is God, here addressing himself only to the temporary needs of the body during its period of test and trial, who does so. But such philosophies offer themselves to be understood not as constitutive of the emergence of science, but as reactive: like Descartes's, they reframe and circumscribe love of the world, but they do not have much success in eliminating it.[36]

If reductionism were the hallmark of seventeenth-century science as signaled in its mechanistic rendering of sensory qualities and in the elimination of scholastic *virtutes*, then one would expect to see this distance from immediate experience, not simply in the posturing of Descartes's fictionalized memoirs, or in Malebranche's proposal that natural science is permissible because perception is the means by which we are intimately linked with God, but in texts and practices generally. But the accusation that the ancients were childish in their sensory superficiality is balanced by the positive valuation, in the educational literature from John Amos Comenius to Locke, of children's naive curiosity and delight in novel experience. Thomas Sprat in his *History of the Royal Society* argues that the formality and confinement of teaching by precepts and universal rules "suppresses the *Genius* of *Learners*," and asks whether "it were not profitable to apply the eyes, and the hands of Children, to see, and to touch all the several kinds of *sensible things*. . . . In a word, Whether a *Mechanical Education* would not excel the *Methodical*?"[37] Even Plato, he says somewhat implausibly, encouraged the hands-on approach. The charm of what Hooke calls the

---

[35] Bacon, preface to *New Organon*, in *Works* 8:26.

[36] Contrast the assumptions of those who, like Susan Bordo in *The Flight to Objectivity*, find themselves methodologically in the intellectualist tradition of which they morally disapprove and read Cartesianism as an alienated philosophy of nature.

[37] Sprat, *History of the Royal Society*, p. 329.

"real, the mechanical, the experimental philosophy" lies in its similarity
to child's play, though a faint aura of sinfulness remains in our "not
only *beholding* and *contemplating*, but . . . tasting too these fruits of
Natural Knowledge, that were never yet forbidden." Hooke can recommend it to gentlemen:

> And I do not only propose this kind of *Experimental Philosophy* as a matter
> of high *rapture* and *delight* of the mind, but even as a *material and sensible
> Pleasure.* So vast is the *variety of Objects* which will come under their Inspec-
> tions, so many *different wayes* there are of *handling* them, so great is the
> *satisfaction* of *finding* out *new things,* that I dare compare the *contentment*
> which they will injoy, not only to that of *contemplation,* but even to that
> which most men prefer of the *very Senses themselves.*[38]

The distance from experience of scholasticism could thus be opposed in
a number of ways: by mystical first-person knowledge (Paracelsianism);
by intuition into foundations (Cartesianism); or by experimental science. Each of these modes of epistemological address invokes a different
notion of immediacy, of which the last finally achieved cultural preeminence.[39]

Reading documents of the period, one is repeatedly struck by references to seeing, and seeing for oneself—to widening the horizon of visual experiences by direct acquaintance with objects or with the help of
pictures and models. As Gillispie says, the anatomy of Vesalius is not
new in content; what is there was already in Galen. But that a revival of
pagan anatomy was taking place is surely important for the development of science, and the desacralization of the body that it implies is
surely relevant to the emergence of scientific objectivity. The new presentations—the large-format depictions of skeletons and musclemen,
the outer layers shown torn away to reveal the interior—were spectacles that commanded interest. The dissection of the human body was
formerly permitted in Christian cultures only to satisfy the requirements
of legal evidence in cases of suspected murder. In Italy, the public anatomy became, in the mid–seventeenth century, a form of public festival:
"nor was it spurned by gentlewomen, who, clad in sumptuous raiment,
attended to the lugubrious exercises of the anatomist during the day,
and later went on to gay balls and parties."[40] The Italian preeminence in

---

[38] Hooke, preface to *Micrographia.*

[39] For a similar analysis, which neglects only the "immediacy" of Cartesian mechanism,
see Blumenberg, *Die Lesbarkeit der Welt,* p. 63.

[40] Adelmann, *Marcello Malpighi* 1:88, quoting Francesco Cavazza, *Le scuolo dell'an-
tico studio Bolognese* (1896), pp. 259–60.

anatomy, which lasted well into the eighteenth century, was a result of this freedom.[41]

The educational writings of Comenius, who, through his disciple Samuel Hartlib in England, was indirectly involved in the formation of the English "philosophical college" of 1645, the precursor of the Royal Society, connect reformed pedagogy and methodical science in the utopian imagination. "Everything should, as far as is possible," he says in his treatise on education, *The Great Didactic* (1657), "be placed before the senses. Everything visible should be brought before the organ of sight, everything audible before that of hearing. . . . the truth and certainty of science depend more on the witness of the senses than on anything else."[42] This means, he continues, "that we must look straight at objects and not squint, for in that case the eyes do not see that at which they look, but rather distort and confuse it. Objects should be placed before the eyes of the student in their true character, and not shrouded in words, metaphors, or hyperbole."[43]

"You then," says Charles Hoole, Comenius's first English translator, in his preface to *Orbis pictus* (1659), "that have the care of little children, do not much trouble their thoughts and clog their memories with bare grammar Rudiments, which to them are harsh in getting, and fluid in retaining; because indeed to them they signifie nothing, but a mere swimming notion of a general term, which they know not what it meaneth, till they comprehend particulars."[44] This message eventually reached Locke, who would turn a normative program for the education of children into a genetic account of the origins of knowledge. We know from Jonathan Swift's parody how readily available as a cultural reference this talk of "things" and of the need to secure direct reference had become.[45] "Those things," Comenius says, "that are placed before the intelligence of the young, must be real things and not the shadows of things. I repeat, they must be *things*; and by the term I mean determinate, real, and useful things that can make an impression on the senses

[41] Cf. de Graaf, "When I travelled through various parts of France for about two years and in many places dead bodies were put at my disposal, I took great delight in opening them and particularly when I set my knife in the Pancreas or the Reproductive Parts, since I daily observed many things in and about these Organs, which had never before been brought to light by any dissector." His work led him, he says, "to be cherished with special affection" by the most learned and inquisitive men of his time. Letter to Sylvius, 20 February 1668, in *De mulierum organis generationi inservientibus*, p. 14.

[42] Comenius, *Great Didactic*, pp. 336–37.

[43] Ibid., p. 338.

[44] Comenius, *Orbis pictus*, p. xxv.

[45] Swift's Laputuans, encountered by Gulliver, stopped talking and only held up objects to secure direct reference.

and on the imagination." Copies, models, and illustrations should be used when the objects themselves are not available, and we need to open things up—just because inside of them are more things. "Botanists, geometricians, zoologists and geographers ... should illustrate their descriptions by engravings of the objects described," for verbal descriptions do not impress the young. To teach anatomy to children one should build a model:

> A skeleton should be procured (whether such a one as is usually kept in universities, or one made of wood), and on this framework should be placed the muscles, sinews, nerves, veins, arteries, as well as the intestines, the lungs, the heart, the diaphragm, and the liver. These should be made of leather and stuffed with wool, and should be of the right size and in the right place, while on each organ should be written its name and function.[46]

A separate memorization of simple parts and a grasp of their relations is essential. In a sentence, "it is often in the smallest words, such as prepositions and conjunctions, that the whole sense depends." And the smallest parts of the body have a similar importance: "Certain it is that in a clock, if one pin be broken or bent, or moved out of its place, the whole machine will stop. Similarly, in a living body, the loss of one organ may cause life to cease."[47] The old arts of memory are turned here to a new use; as a grown man can master and retain in visual memory the contents of a whole palace, "just as easy will it be for a youth who is admitted to the theatre of this world to penetrate with his mental vision the secrets of nature and from that time forward to move among the works of God and man with his eyes open."[48] But the possibility of universal knowledge, which Comenius derives from the Bacon of the *New Atlantis*, no longer implies the necessity of holding everything in one's head at once. Images, models, and illustrations serve a purpose both decorative and utilitarian in the ideal cities of utopists. In Tommaso Campanella's "city of the sun," Wisdom, one of three collateral princes, has charge of all the sciences, which he teaches to people as Pythagoras did to his disciples. There are pictures of all the sciences painted on the circular walls of different segments of the city: pictures of stars, countries, alphabets, mathematical figures, minerals, metals, stones, lakes, seas, rivers, wines, oils, and herbs, together with written descriptions of how plants, stars, and metals are related to parts of the body and to medicine. There are displays of samples of stone and carafes of liquids, and more representations—of fishes, birds, quadrupeds, and insects. "All the mechanical arts are displayed, together with their

[46] Comenius, *Great Didactic*, p. 338.
[47] Ibid., p. 343.
[48] Ibid., pp. 344–45.

inventors, their diverse forms and their diverse uses." And thus "without effort, merely while playing," the children of the solarians "can come to know all the sciences pictorially before they are ten years old."[49]

Though there were points of connection, the pansophists taught a message contrary to the allegedly Cartesian one of reduction and abstraction. "The eye," Comenius says, "naturally thirsting for light, . . . rejoices to be fed by gazing, and suffices for all objects, . . . and just as it can never be satiated by seeing, so does the mind thirst for objects, ever longs and yearns to observe, grasps at, nay seizes on all information and is indefatigable."[50] The mind is like a spherical mirror suspended in a room, which "reflects images of all things that are around it."[51] The world is "a garden, a theater . . . to be seen, heard, smelt, tasted, handled." Visual perception produces an "abyss" of images, compounded and multiplied daily.[52] To teach is to indulge the craving for facts, not to stuff and flog the mind with the "husks of words and chaff of opinions," which make the pupil fastidious and exotic; all of Comenius's love for youth "who have been so much vexed" is apparent here. The teacher is to lead the pupil to see the world whole, plainly and lucidly, not "obscurely, perplexedly, and intricately, as if it were a complicated riddle."[53]

The programmatic statements of the seventeenth-century atomists and corpuscularians are, then, not the meaningful figure against the background of a proliferating but inchoate empiricism; they are philosophy, which is to say that they are not only, and perhaps not primarily, a support for science, but also a reaction to it and a substitute for it. Attempts to capture the sensory richness of protoscience, its concreteness, its disorderliness, its distaste for abstraction and systematization, its agnosticism on questions of basic ontology, and its suspicion of great theoreticians, must depend to some extent on the record of what has not survived to become incorporated into modern science. Lynn Thorndyke's digests of the transaction books of early scientific societies and their published ephemera catch scientific practice in some of its least self-conscious, least artificial, most representative moments: we see the scientific revolution in dishabille, as it were.[54] But Thorndyke's approach was, on the whole, rather ironic. My purpose here is the more earnest one of habilitation, a task that demands that we do not ask the period to be something other than it is. The introduction of quantifica-

[49] Campanella, City of the Sun, pp. 36–37.
[50] Comenius, Great Didactic, p. 198.
[51] Ibid., p. 193.
[52] Ibid., pp. 194–97.
[53] Ibid., p. 230.
[54] Thorndyke, History of Magic, vols. 5–7.

tion into the qualitative sciences, which Westfall saw as frustrated by an unresolved tension or conflict, later somehow released, needs a different interpretation. In anatomy, medicine, and embryology, the first movements toward quantification appear to have less to do with a mystical Pythagorean impulse to express everything in numerical terms, thereby getting closer to reality, than with fussy habits of counting, timing, calculating, and measuring.[55]

The rhetoric of new beginnings and new methodological foundations can obscure the ways in which seventeenth-century science was a restoration and a continuation of the reasoned natural history of the ancients, which had been lost in, and to a certain extent repressed by, Christian culture and scholastic philosophy. The frequent harping on the errors of the ancient naturalists does not signify simply a break with the past, but a continuing involvement with it; the moderns were equally critical of one another's errors and found sufficient provocation in them. The Aristotle who had written about the chicken's egg was acknowledged to have been wrong in much of what he said, and he had failed to detect very much happening in the first days of incubation. But there was nothing in his attitude toward the study of the egg that needed correction: the thing simply had to be done better, more precisely. One reason it could now be done better was that the technologies of reading, writing, and looking were now entirely different. Success in drawing in and equipping new observers and theoreticians of nature, in creating a scientific community, depended on the introduction of print and was assisted by the development of techniques for reproducing visual images in printed form.[56] Anatomical works, letters and reports to scientific societies, specially commissioned studies, compendia, and review journals, not philosophical treatises, were what created a scientific community.

There had been no lack of scientific talent in the Hellenistic world; there was no perversion of the intellect created by the numbed acceptance of Aristotelian metaphysical dogmas that inhibited the development of science, and, despite the lack of encouragement of the medieval church for the exploration of unknown realms, the medieval mind was not, because of a failure to conceptualize sufficiently abstractly, incapable of natural science either. True, Aristotle had set loose, against the free speculations of the mechanists and atomists that referred to a nonvisible underlying reality, the philosophical weapons of distinctions in language and appeals to experience. The association of atomism, mech-

[55] See Heilbron, "Measure of Enlightenment."
[56] See Eisenstein, *Printing Press*, esp. part 3; Ivins, *Prints and Visual Communication*.

anism, and atheism in Lucretius discredited the search for natural causes and impugned the belief in natural necessity. But there is an element of contingency here. Aristotle was cherished because he was there: had Democritus's writings been as comprehensive, and had they survived in manuscripts and translations as Aristotle's did, to be gathered up in printed collections in the fifteenth and sixteenth centuries, Christianity would no doubt have found its reconcilers and made its peace with atomism. Had Augustine not triumphed over his ecclesiastical rivals, the inward orientation of Christianity might not have been so pronounced, its fear of demons so exaggerated; had its positive aspects not been structured around miracles and transformations, Christianity's distrust of rival magic might have been less sharp.[57] But without its texts, science would have fared no better. The multiplicity of print, as Elizabeth Eisenstein has argued, stimulates comparison, comparison stimulates criticism and revision, revision stimulates progress, and all of this stimulates imitation.[58]

This is not to say that philosophy and reconceptualization had no role to play. Print brought back, in addition to atomism and mechanism, Greek rationalism, which insisted on natural causes for all phenomena, and which was finally made more convincing even to Christian natural philosophers than the orthodox view of the specificity and intentionality of divine action. But what made the doctrines of the ancient atomists so profoundly attractive in the early modern period was not their intellectual economy, austerity, and remoteness from experience. These characteristics remained, as they had been for Aristotle, drawbacks. Their attraction lay rather in the hope of rendering the causes of phenomena perspicuous, which depended in turn on the hope that the limits of visibility could be pushed back indefinitely far. Here the development and deployment of optical instruments had a critical role to play. And it was through instruments as well that the observational science of the ancients was converted into experimental science, the gathering of information through the generation of artificial situations and artificial perspectives.

Attention to the literary dimension, then, furnishes only part of the story of the emergence first of protoscience and later of science. Equally important were their loci and their props. The laboratory, an alternative site of intellectual labor to the solitary study of the humanist scholar, appeared both in the utopian imagination and in fact in the first half of the century. Well before the foundation of the English Royal Society, the Italians had established their own societies for demonstration and

---

[57] On magic and the church, see Walker, *Spiritual and Demonic Magic*.
[58] Eisenstein, *Printing Press* 1:74ff.

experimentation. These included the Lyncean Academy, founded by the duke of Cesi, which conducted some of the first microscopical observations and which survived from 1603 to 1629; and the quarrel-ridden Academy of Cimento, which lasted only ten years after 1657, whose members worked on measuring, freezing, atmospheric pressure, experiments with metals and magnetism, the compression of water, firearms, and natural history.[59] The Royal Society, which had existed as an informal group since 1645, was chartered in 1662, and the French Académie des Sciences soon thereafter, in 1666; of the German societies, the most famous was that of J. C. Sturm, the Collegium Curiosae, which met in Nuremberg, the site of trade in toys and instruments.

The laboratory or the demonstration room of an early scientific society was not only a modification of the scholar's study, but the extension of other places: the workroom and the nursery with their tools and playthings, the anatomical theater of the medical school, the autopsy room where the causes of death were investigated, the astronomical observatory where data were recorded, and the collector's cabinet, which was itself a miniaturized world of samples. The participants were at once collectors, spectators, and actors. Their role was to bring nature, and not only nature, but geography, manufacturing, and tourism, within doors, to sweep as much as they could of the world into their regular meetings. They might be relatively passive with respect to it, admiring, recording, and classifying, or active in their experimental interventions—their suffocations, poisonings, transfusions, tooth transplants, and so on. This creation of a copy of the world was, in the Augustinian sense, concupiscence, a desire for possession,[60] just as there was something demonic in the favored interventions into organic life of these early experimentalists. But it was from this nidus that the methods and the knowledge of science emerged.

The desire to have one of everything, to possess the whole world in microcosm, to learn from things themselves without the medium of arguments and disputations, informed both the language and the practice of the early scientific societies and the informal groups that preceded them. There is a strong similarity between the early accounts given in the *Journal Books* of the Royal Society, which recorded events of the weekly meetings, and the cabinets of the early collectors: the juxtaposition of what are from our point of view unrelated elements contrasts oddly with the methodicality and solemnity of these occasions and these

[59] See Middleton, *Experimenters*, pp. 7ff.

[60] Cf., however, Lewis Feuer's hard-to-assess claim that science was the creation of hedonists reacting against Puritan ascetics (and against feminine masochist-ascetic Aristotelians); *Scientific Intellectual*, p. 54.

collections. When the king's councillor Balthasar de Monconys traveled through England, France, the Low Countries, Spain, and Egypt in the middle of the century, he was asked to admire and sometimes to pay for the privilege of viewing private museums, with their paintings of the nobility, their surgical and mathematical instruments, their ostrich eggs, crystal vases, coconuts, and automata.[61] Why exactly these objects and not others? At best it can be said that each was in some way rare, scarce, ingenious, odd, or distinguished: otherwise they had nothing in common. In the museum of Mr. John Tradescant, runs an account written in 1638, were the following things:

> first in the courtyard there lie two ribs of a whale, also a very ingenious little boat of bark; then in the garden all kinds of foreign plants. . . . In the museum itself we saw a salamander, a chameleon, a pelican, a lanhado [snake] from Africa, a white partridge, a goose which has grown in Scotland on a tree . . . a number of things changed into stone, amongst others a piece of human flesh on a bone, gourds, olives, a piece of wood, an ape's head, a cheese, etc.; all kinds of shells, the hand of a mermaid, the hand of a mummy, a very natural wax hand under glass, all kinds of precious stones, coins, a picture wrought in feathers . . . the passion of Christ carved very daintily on a plumstone, a large magnet . . . a scourge with which Charles V is said to have scourged himself, a hat band of snake bones.[62]

Elias Ashmole's catalogue of the Tradescant collection made in 1656 is only somewhat more systematic. It includes petrified objects, both manmade and natural; miniature carvings; and other products of art and nature: castings, moldings, metals, earths, salts, exotic clothing, and accessories.[63] It is difficult to distinguish the real collection from the collection bequeathed by Sir Nicholas Gimcrack in Thomas Shadwell's parody *The Virtuoso*.[64] And it resembles in overall texture the ceremonies of the Royal Society reported in the *Journal Books* for 1665–66, in which the making of hats, viper powder, and the inspection of small glass balls are considered in rapid succession. The antiquarian virtuoso had become a scientific virtuoso: not a collector of pagan relics, but a student of the creation. Curiosity was still, however, the dominant motive.[65]

It is not surprising that the question of the usefulness of knowledge of nature kept arising in the second half of the century, and the Royal Society was often trying to reassure itself on this point. In the "Grand

[61] Monconys, *Journal des voyages* 2:247.
[62] Gunther, *Early Science in Oxford* 3:284.
[63] Ibid., pp. 436ff.
[64] See Marjorie Nicolson's introduction to Shadwell, *Virtuoso*, p. xvii.
[65] See Lloyd, "Shadwell and the Virtuosi."

Academy of Lago" encountered by Swift's Gulliver in the course of his travels, such profitable experiments as extracting sunbeams from cucumbers, calcining gunpowder, reducing excrement to food, and breeding spiders for the silk trade are underway. The dialectic of frivolity and profit is resolved in increasingly conventional appeals. The preface to the catalogue of the Pointer Collection of 1740 says that although its objects may appear inconsiderable in the eyes of the vulgar and illiterate, they all have their separate "Use and peculiar Advantage": the artifacts teach history; the plants, seeds, earths, metals, and minerals are of use in medicine and mechanics; everything else in the way of natural wonders teaches the power and art of God. The Royal Society's interest in manufacturing processes, navigational aids, and agriculture might have seemed easy to defend as economically significant. But Thomas Sprat had to argue that this was so, and that the mechanical arts were not, as was commonly believed, incapable of further improvement or subject to improvement only by the chance inventions of artisans themselves. Like Descartes, he tried to push for rational changes in technique, as opposed to those dictated by arbitrary changes of fashion. When manufacture is subject to scientific control, he urges, "the weak minds of the *Artists* themselves will be strengthen'd, their low conceptions advanc'd, and the obscurity of their shops inlighten'd . . . the flegmatick imaginations of men of *Trade*, which use to grovel too much on the ground, will be exalted." And, correspondingly, "the conceptions of men of *Knowledge*, which are wont to soar too high, will be made to descend into our *material World*."[66] Yet although the early Royal Society collected information from artisans and mechanics—William Petty's treatise on dyeing and coloring furnishing the best example—instrument making itself rather than any industrial process provides the best example of attempted exchanges between workers and theoreticians.

A certain humor is apparent in the *Transactions* of the society for 1676, which contain, in addition to abstracts of mathematical treatises, accounts of tours in Greece and Italy, a notice concerning eclipses, and the review of a book on scurvy grass that "will instruct all sorts of persons, how to make Wines, Sauces, Syrups, and distill'd waters of this Plant, for the good of their sick and languishing neighbors."[67] Tourism was not considered idleness and aestheticism in the seventeenth century, but a form of industry, the collection of the practical knowledge and manufacturing expertise possessed by people of other countries. But what then is the relation between science and the promiscuous collecting, the increasingly tired straining after uses that, after the promises of

---

[66] Sprat, *History of the Royal Society*, p. 396.
[67] Review of "An Account of *Cochlearia Curiosa*," p. 621.

Bacon that nature would be mastered and redirected by the imposition of methodicality and group effort, threatened to deteriorate into vacuous amusement? One view is that protoscience is not science. The activities of the early societies were not, apart from the presence of a few presentations and papers involving geometrical or numerical reasoning and systematization, scientific. Had Newton, in all his personal and intellectual austerity, not assumed its presidency in 1703, the Royal Society, which became as a result a less erratic and luxurious enterprise, would not have survived. Savage criticism and lighthearted parody might well have exterminated it between them. William Wotton expressed this worry in 1694: "Howe far this [ridicule] may deaden the industry of the philosophers of the next age is not easie to tell."[68]

But the statement that protoscience was only a precursor of science leaves us with the following puzzle. The *Journal Books* of the Royal Society furnish us with a weekly account of what was said and done in the society's meetings. Let us concede that what was said and done between 1665 and 1675 was predominantly unscientific by modern standards: that the interest was antiquarian or virtuosic or amateurish. Nevertheless, there was a possible route from this protoscientific pattern of activity to a scientific one, because there was an actual one. There is no moment—no weekly meeting—at which one can say that the activities of the society became scientific, and remained so. It is thus important to ask what in the protoscientific period of the society made the direction it eventually took possible. Recently Lorraine Daston has addressed this question, arguing that the attempts by early academicians to study "facts" and events independent of any systematization are precisely what provide the link. It was in the "fragmentary syntax" of the collection and the meeting that objectivity began to take hold. Curiosities, monsters, and isolated examples disrupted inherited theories about the regularity of nature; they forced a reassessment and a reopening of the entire subject of natural forces and natural virtues.[69]

This suggestion may eventually furnish a large piece of the puzzle of the connection between protoscience and our science. In the meantime, in trying to understand this relation, it is unhelpful to distinguish between scientists and amateurs, to import a distinction that, unlike the distinction between virtuosi and tradespeople, was not available to the participants themselves. Certainly some participants and correspondents, such as William Croone, Kenelm Digby, and Athanasius Kircher, were regarded as less reliable and more given to whims and fancies than

---

[68] William Wotton, *Reflections on Ancient and Modern Learning*, in Stimson, *Scientists and Amateurs*, p. 71.
[69] Daston, "Baconian Facts."

others. But they were not regarded as nonparticipants in some enterprise that excluded them and all of their ilk. It is correct but somewhat misleading to say of the microscopist Leeuwenhoek, as Marjorie Nicolson does, that he was "without scientific training, in no sense a philosopher, not even a learned man, as the century understood learning."[70] Humanistic or philosophical education would not have brought Leeuwenhoek farther than he came, and the very concept of scientific training had no application in the first period of microscopical research.

The "seriousness" of those whom we can identify as the founders of biology and physiology cannot consistently be referred to their institutional training or professional positions. Remarkably, it shows itself in their habits of study; they can be recognized by their single-mindedness, their refusal to be drawn off the track, their rejection of the temptation to polymathy. The difference in mental organization between a Leeuwenhoek, a Jan Swammerdam, a Malpighi, and the ordinary amateur is a matter of application and thoroughness. But the soil of empiricism and sensualism is, for all that, the same soil. Nor is it obvious that the scientific spirit is present in Malpighi, absent in his medical-school critics, present in Francesco Redi, who attacks generation from putrefaction, and absent in Kircher, who defends it and palingenesis too, present in the Royal Society, absent in Swift and Shadwell, who mock it, and absent in Henry Stubbe, who argues that scientific societies are both inept and irreligious.

There is no historical moment at which the virtuoso was no more, having been succeeded by two new kinds of being, collecting dilettantes and dabblers who amused themselves on one side, methodical practitioners who observed experimental protocol and supplied results on the other. The categories of amateur and professional emerged gradually through the creation of institutions for classifying and sorting results, perpetuating knowledge, and training students in the management as well the production of information. The descendants of Descartes comprise both theoreticians and fantasists; but Descartes himself is neither one nor the other, for there was no structure in 1640 that could prise and hold those categories apart. The language of sobriety and plodding methodicality, the scorn for the imaginary, was there; but it was translated into fact only through a slow process that occurred as organizational practices and styles of individual activity were discarded and imitated. This process did not occur through the agency of some selecting *Geist*—the new scientific spirit of the age—but as organizations and individuals made particular choices: what to pay attention to, what to

[70] Nicolson, *Science and Imagination*, p. 165.

publish, what to copy, whom to be like. "Our company of philosophers thinks," Henry Oldenburg, the secretary of the Royal Society, wrote to Malpighi in 1669, "that you are treading the real paths leading to a true knowledge of nature's secret places when you put aside the empty arguments of the schools' theory for pursuing almost nothing but generalities, and devote your mind and hands to observing accurately and eviscerating minutely the things themselves."[71] These gentle acts of praise and helpful offers of publication, repeated many times over, ensured, in the long run, the dominance of certain scientific personalities, certain modes of presentation.

The cliché about eighteenth-century science is that it is "taxonomic," and hence trivial, or at best propaedeutic. The implication is that the naming and describing of plants, animals, diseases, or chemical substances on the basis of their phenomenological qualities is inferior to the bold reductions, the extraordinary revisions, of the philosophical minds of the seventeenth century. The more confident among them believed in the capacity of the student of nature to arrive at a knowledge of real essences, not the nominal ones that permit classification from the human point of view. Are not sorting and collecting by contrast the pastimes of children, or even of rats and crows? A mature science, a profound science, asserts the existence of invisible structures and relations, which knowledge it can arrive at only by indirect means.

One may wonder whether the eighteenth century was really less speculative by any intelligent measure than the seventeenth, not merely increasingly comprehensive and familiar in its classifications. But a more serious criticism of the accusation is that there are not two exclusive procedures, that of reconstituting the world in microcosm by collecting everything and writing down everything that has been observed to happen, and that of projecting views of the fundamental structure of the world arrived at by immediate intuition. As Ian Hacking has argued, the true dialectic is that which takes place between representation and intervention.[72] Both collecting and theorizing are, in the end, only forms of representation. Each alone is powerless to give substance or movement to science, but they are equally powerless in tandem. Experimentation too, if suitably ritualized, may become itself only representation or display: *Provando et riprovando*—test and retest—was already the motto of the Academy of Cimento, which was more interested in repeating striking demonstrations than in promoting or settling contro-

---

[71] Letter to Malpighi, 25 March 1669, in Adelmann, *Marcello Malpighi* 1:674.
[72] Hacking, *Representing and Intervening*, esp. pp. 262ff.

versies, or in evolving theories, or in establishing priorities for investigation.[73] But there is perhaps something unsatisfactory about the story in
which the conceptual profundities of seventeenth-century science are
followed by the taxonomic trivialities of the eighteenth. Perhaps it is in
part a case of *reculer pour mieux sauter*: proper classification is the dull
task that precedes theoretical progress; the underlying work must be
done if theorizing is not to be premature and ill-founded. As Bacon
said, first you catalogue, then you sift, and only then do you attempt, by
suitable applications of intellectual machinery, to arrive at a knowledge
of invisible structure. A collective, somewhat belated, realization of this
notion might thus help explain this supposedly regressive movement.
But reactions against the prematurity of profound science and the triviality of a science of collection and classification were already well established in the seventeenth century, and the taxonomic drive of the eighteenth century was in many ways simply an extension of the ventures of
early collectors and displayers, which can already be seen in the patient
and exacting classification of fish and plants in, for example, the books
of John Ray, or in Swammerdam's insect museum. The bold theorizing
of the seventeenth century never intersected at all with the anatomizing
and naming, the promiscuous collection; the perceived tension here is as
imaginary as that between mathematics and micromodels.

Experimentation is, however, neither theoretical representation, nor
taxonomy, nor the mere display of phenomena.[74] And for this reason it
is precisely the aimless-seeming experiments of the early Royal Society
that are the first steps toward the exact protocols and the aggressive
creation of unnatural states of affairs that distinguish science from protoscience. We still lack a general study of early conceptions of experimental protocol and experimental performance, report, and dissemination, a study that would add considerably to our understanding of how
modern science came to be. Such a study would be able to embrace the
content of works as diverse as Francesco Redi's *Experiments with Vi-
pers*, in which the existential interest in poisonous snakes gives rise to a
bevy of procedures and operations that an external anthropological observer of seventeenth-century European science would be hard pressed
to see as leading to new knowledge, and Newton's *Opticks*, or the now
better understood set of manipulations involving the air pump.[75] Considered under the aspect of representation, these manipulations were

[73] See Middleton, *Experimenters*, pp. 52–53.

[74] As Peter Dear notes, the production of artificial experiences is conceptually difficult;
they belong neither in the category of events belonging to the regular course of nature,
which is what we are trying to understand, nor in the category of the monstrous. "Jesuit
Mathematical Science," p. 160.

[75] See Shapin and Schaffer, *Leviathan and the Air-Pump*, esp. chap. 2.

simply additions to the range of observational experience of the participants: experiential novelties. Even as such, they helped establish the
sense of that powerful but elusive concept of objectivity. But seen under
the aspect of intervention they accomplished something different; they
unearthed what there is in the content of science that is cumulative, that
does not change, or does not change significantly for practitioners, even
when the explanatory theory surrounding them changes.

We need to distinguish between the objectivity of reductive ontologies
or metaphysics and the objectivity of modern science. The former is not
constitutive of and is largely irrelevant to the latter. The experimental
and observational sciences of the seventeenth and eighteenth centuries
were not detached from human interest, from religious meaning, from
an immersion in the density of qualities. But they began to make themselves objective in their constitution of a scientific object: an object that
is collected, labeled, put in a museum, that is sliced, dissected, solidified,
dyed, and put under a microscope. The "spectacle of nature," a concept
popular in eighteenth-century science, denotes something Janus-faced;
the term refers at once to a harmony, a theodicy, a moral lesson, and
something existing neutrally in and of itself.

There is nevertheless something important in the traditional conception of the scientific revolution. First, there was a sense in which the
flight from theory and foundationalism in experimental sciences was
reactive, a frightened withdrawal from "free and Democritean" speculation. The compiling of data and amassing of facts, the performance of
inconclusive experiments and pleasing demonstrations, was unlikely to
provoke an official and hostile reaction because it stayed far away from
the theologically charged questions of ontology.[76] A repetition of the
Galileo affair was ruled out by the innocuousness of much academy
science. Descartes's own withdrawal of his foundational treatise *Le
monde* in 1630, and his presentation of treatises devoted to specific
experimental or practical topics (geometry, optics, "meteors") of no
theological significance, may reflect this substitution of aims.

Second, the idea that science destroys the image of the familiar world
and substitutes for it the image of a strange one, wonderful to the imagination and at the same time resistant to the projection of human
values, is essentially right. The paradox that understanding seems to
imply alienation from what is understood is one presented in the intellectualist version of the scientific revolution, according to which a color-

[76] Thus Leonardo Olschki, "An immense amount of erudition piled up in every corner
of the [Italian] peninsula between 1650 and the Napoleonic era . . . a dead heritage of
those generations" that "still commemorates, in miles of pigskin volumes in provincial
Italian libraries, the unflinching devotion and silent enthusiasm that inspired their authors
and readers." *The Genius of Italy*, in Adelmann, *Marcello Malpighi* 1:119.

less world of particles was given to us in place of the world of experience, this theft sweetened by the knowledge that we had been divested of the illusions of childhood, our naive animism and anthropomorphism, and had achieved a vision of things *sub specie aeternitatis*. For general cultural reasons, which perhaps have more to do with the role of technology in producing alienation and oppression than with the study of texts and authors, this historiography of science has fallen into disrepute, together with its heroes and its values. My intention here is not to minimize the importance of those reasons, but neither is it to join the chorus of those who condemn the early modern period, together with its historians, as a manifestation of alienation and a will to dominate.[77] Certainly we must recognize that the concepts and values of progress and objectivity have become snares for the unwary. And the logic of the concepts involved is such that we are prevented from saying that the nonobjective—the products of illusion and interest, the pathological and the ideological—are themselves given either objectively or else only in those acts of naming that reflect any local and contingent imbalance in power relations.

Without, then, undertaking the impossible and misguided task of defending objectivity, I will be content to present early modern science under a more benign aspect than has recently been customary, showing how engagement exists alongside detachment, and receptivity along with a desire for mastery. Hooke's image of the microscopist catching a glimpse of nature working undisturbed, in its usual way, "through these delicate and pellucid integuments of the bodies of Insects" is, in this respect, a memorable one:

> when we endeavor to pry into her secrets by breaking open the doors upon her, and dissecting and mangling creatures whil'st there is life yet within them, we find her indeed at work, but put into such disorder by the violence offered, . . . it may easily be imagined, how differing a thing we should find, if we could, as we can with a *Microscope*, in these smaller creatures, quietly peep in at the windows, without frightening her out of her usual bays.[78]

That in the very charm of the miniature and the attitude of passivity that observational science inculcated there might be stimulus and encouragement to illusion is a question I will address below.

---

[77] See on Bacon's imagery—and for a strong interpretation of its influence—Merchant, *Death of Nature*, pp. 168ff.

[78] Hooke, *Micrographia*, p. 186; see Fournier, "Fabric of Life," p. 57. As Fournier points out, early experimentalists were not immune from uncertainty and guilt over the suffering to animals caused by their interventions, and Robert Hooke turned to the innocent science of the microscope with some relief.

2

The Subtlety of Nature

IN 1664, the last year of his life, the traveler and microscope aficionado Balthasar de Monconys made some observations on the sedimentation or crystallization of medicine in a glass on his bedside table "with great pleasure, and [decided] that the faculty this medicine had of cleansing the stomach could reasonably be ascribed to the figure of the particles which composed it."[1]

The interest of the anecdote lies in the confusion it implies between observation and theory. Monconys's account of the drug's medical efficacy, an importation from what he had read or heard, makes reference to the visualizable without constituting in any manner a deduction from what he saw. He is not dogmatic—he thinks only that the cleansing properties of the drug could reasonably be ascribed to the shape of its particles. But he does not consider this a matter of pure speculation: he sees the contents of the glass by his bedside as probably composed of subvisible particles that act on the body by contact and pressure. And why anyone looking at a glass of medicine in 1664 should see things in this way requires some explanation, beyond the mere availability of corpuscularian theory. The assertion that the specific powers and perceptible properties of natural substances depend on the combined actions of homogeneous or minimally differentiated material particles lying beneath the threshold of normal perception is the main feature that distinguishes seventeenth-century matter-theory from the Aristotelian theory of substances and qualities. And that the body is an ensemble of micro-machines that can be in principle adjusted by the physician is arguably the most directly influential of Descartes's teachings. Both points—that the body is a machine and that drugs work in it by contact and impact—are made in Robert Boyle's paper "On the Reconcileableness of Specific Medicines to the Corpuscular Philosophy,"[2] and indeed throughout his repetitive and influential exposition of the corpuscularian philosophy, which he sometimes called the "mechanical hypothesis." But the success of this doctrine in the mid–seventeenth century requires more explanation that it is usually given. The account of nature that the corpuscularians proposed against the Aristotelians succeeded, if

[1] Monconys, *Journal des voyages* 2:228.
[2] In *Works* (1772 ed.), 4:301–25.

the standard story is to be believed, only through an intellectual advantage: its little machines could be clearly and distinctly conceptualized. Aristotle's system of qualities, virtues, powers, forms, and potentials was not so agreeable in this respect. But this account, I believe, puts the cart before the horse. There is reason to doubt that these conceptual advantages—simplicity, economy, universality, and so forth—that figured so prominently in the rhetoric of the moderns, actually explain the success of the corpuscularian philosophy in winning general acceptance. Despite the tendency to present a seventeenth-century Democritean revival in these terms, there was no sudden upset of the old ontology of substances, manifest and occult qualities, virtues, and forces, and its replacement with an incommensurable one. Rather, the corpuscularian philosophy established itself in the first half of the 1600s as the product of a progressive refinement of the Renaissance notion of "subtlety," and a materialization of hidden resident spirits. And here the microscope, as Monconys's account suggests, had a role to play.

"My purpose in writing this book," Girolamo Cardano states in the first book of his *De subtilitate* of 1560, "is to expound the significance of *subtilitas*." What is *subtilitas*? It is, he says, "a certain intellectual process whereby sensible things are perceived with the senses and intelligible things are comprehended by the intellect, but with difficulty."[3] The book is a discussion of those things that can only be known with difficulty because they are neither clearly perceivable by us nor clearly thinkable: "If obscurity begets difficulty, this book chooses for discussion only the most obscure subjects."[4] The difficulty he has in talking about the subject is thus self-explanatory. But by way of clarification, he tells us there are two main types of obscurity. "Although obscure things are for the most part small and small things elusive, yet not all things are so nor have they been so always. For the things which through impropriety of language rather than intention are unintelligible and entangled like knots which chance to twist upon themselves . . . do not deserve the name of *subtilitas*." They resemble rather the withered legs of starved or crippled men. But is subtlety not, he asks, something absurd, something superfluous—like the flea on a golden chain or the tiny flies modeled by Callicrates in ivory? No: it is only the subtlety of human art, he says, that is so useless; the subtlety of nature has in it something of the celestial.[5]

What exactly is Cardano talking about? What he says here amounts

[3] Cardano, *First Book of Jerome Cardan's De subtilitate*, p. 75.
[4] Ibid.
[5] Cardano, *De subtilitate*, book 12, in *Opera omnia* 3:364.

to a declaration—against the spirit of Aristotle—that the truth about things is not reached by ordinary processes of reasoning and observing. Like Nicholas Cusanus, who, in his philosophical treatise on origins and enigmas, *De beryllo* (1488), compares the magnifying lens of the mind to the colorless stone that renders the invisible visible, Cardano pairs intellectual and visual subtlety. "The subtlety of nature," Bacon would say, helping this thought evolve into a comparison between nature's productive faculties and our cognitive ones, "is greater many times over than the subtlety of the senses and understanding."[6] Modern epistemology is characterized, and distinguished from ancient epistemology, by its exploration of match and mismatch in this respect.

The recalibration of human knowledge with respect to the very small antedates the development of mechanical philosophy, considered both as an ontology and as a set of experimental programs. We can hear the echoes of Cardano, mediated through Bacon, even in Hooke: the search for causes is hampered, Hooke says, not only by the difficulty, obscurity, and sensory inaccessibility of the material we work on, but also by the obscurities of language: "'tis no wonder, that our power over natural causes and effects is so slowly improv'd, seeing we are not only to contend with the obscurity and *difficulty of the things* whereon we work and think, but even the *forces of our own minds* conspire to betray us." The superiority of what he calls the "*real*, the *mechanical*, the *experimental* Philosophy" over the "Philosophy of *discourse* and *disputation*" is that the former does not aim at the "subtilty of its Deductions and Conclusions" but at the proper establishment of groundwork, this to be accomplished by a "*watchfulness over the failings* and an *inlargement of the dominion*, of the Senses."[7] The difference is that the subtleties of art and human manufacture prove to be adequate to the task. The microscope, a subtlety of manufacture, and the subtleties of nature it revealed were the means by which a Baconian theory of the interpretation of nature was enabled to take hold and drive out its rivals, bypassing the obscurities of language and offering a resolution of visual confusion.

Its rivals were, even at midcentury, numerous. John Webster, the Puritan divine who, in his *Academiarum examen* of 1654, mounted an attack on the universities to force them to replace the "rotten and ruinous fabrick of Aristotle and Ptolemy," was ready to substitute for them any and all of Bacon, Robert Fludd, Oswald Croll, Gassendi, Kepler, and Descartes. Like other reformers who opposed scholastic and humanistic culture, Webster knew what he did not like: "grammar," by

---

[6] Bacon, *New Organon* 1:x, in *Works* 8:69.
[7] Hooke, *Micrographia*, preface.

which he apparently meant not only the structure of languages but also formal reasoning and disputation, logic, and rhetoric. What he approved of, engagement with nonlinguistic reality, suggests empiricism, but it is empiricism with a difference. In place of the school curriculum, he recommends that "probable, pleasant, and useful . . . Hieroglyphic, Emblematic, Symbolic and Cryptographical Learning" allegedly promoted by his heroes.[8] Webster contrasts the "dead paper idols of creaturely-invented letters" with the "legible characters that are only written and impressed by the finger of the Almighty," a Christian philosophy of created things, in contrast to the pagan philosophy of heathen letters. The schools, he says, are ignorant of physiognomical science, which alone "doth explicate the internal nature and qualities of natural bodies" and so teaches us "all the salutary and morbidifick lineaments."[9] Living things are "not as mute statues, but as living and speaking pictures, not as dead letters but as Preaching Symbols."[10] And he goes on to contrast the "vulgar anatomy and dissection of the dead bodies of men" practiced by William Harvey with the "vive and Mystick Anatomy" that alone can penetrate to the causes and cures of disease.[11]

The defense of the universities was undertaken promptly by the mathematician and astronomer Seth Ward and the encyclopedist and language-projector John Wilkins, who focused their attack on "that canting Discourse about the language of Nature." Webster, they charged, was incapable of understanding the variety and elementary nature of signs. He could not tell the difference between the mathematical innovations of the algebraist Vieta and charms and magic formulas. And there is no use in studying hieroglyphs if one wants to achieve insight into the secrets of nature; nature is not an enigma to be unraveled or a symbol to be decoded. Webster's undiscriminating references to all self-professed reformers showed that he had understood none of them; had he understood even one of them well, he could not have incorporated them into a single category. "There are not two waies in the whole World more opposite," Wilkins and Ward state, "than those of L. Verulam

---

[8] Webster, *Academiarum examen*, p. 24. Here Webster seems to be retreating somewhat from the position of Robert Fludd, who in his own 1617 polemic against academic learning distinguished between the Egyptian hieroglyphics "whose letters, images, and characters are artificial, superficial, like shadows of real things and the products of human invention," and the things of nature, the "celestial hieroglyphs," which are not "superficial, imaginary or dead but on the contrary, substantial, real and living." Fludd, *Tractatus apologeticus*, p. 45.

[9] Webster, *Academiarum examen*, p. 74.

[10] Ibid., p. 28.

[11] Ibid., p. 74.

and R. Fludd": one is a philosophy based on experiment, the other on "Mystical Ideal Reasons."[12]

According to Wilkins and Ward, although both Bacon and the occult philosophers had effectively set aside grammar and disputation, there was a good deal to choose between them. The allure of Fludd, with his macrocosmic-microcosmic correspondences and musical-celestial analogies, and Croll, with his theory of signatures, was still strong. The occult philosophy of the Renaissance, unlike the physics and metaphysics of the schools, did project a horizon of new knowledge before it— knowledge that would be spiritually ennobling as well as useful—and it carried with it a utopian and antiauthoritarian political message that was welcome particularly in England during the civil war and interregnum years. The philosophy of Paracelsus and his followers was revived in the 1640s and 1650s, with a great outpouring of occult literature from the press;[13] and it is important not to underestimate the degree of engagement with it on the part of the orthodox. Mersenne had written a polemic earlier in the century against magic and the qualitative nature philosophy of Campanella; Kepler and Gassendi wrote against Fludd; and Descartes referred obliquely to an early fascination followed by a disgust with magic. It is easy to imagine philosophical rationalism, manifested in methodologies employing new and stringent canons of evidence, driving out irrationalism, as though the *Geist* of history were at work. It is less easy to see how, in terms of the choices actually available, certain modes of practice and language came to achieve authority over others. One way to understand this is by looking at how the belief in hidden powers, in the subtlety of nature in Cardano's sense, and in the mandate of the occult philosophy to seek them out was demystified and attached to concrete and specific procedures of inquiry, creating, in an inchoate antischolastic nature philosophy, an eventual division between science and nonscience.

According to Paracelsian doctrine, the powers and forces of things reside in them, as spirits live in bodies, and as odors, which may be released by grinding and crushing, or captured by concentrating and distilling, reside in herbs or in chemical mixtures. These spirits have noxious or curative powers that are of great importance to us, but they are hidden from the bodily eye. Only spiritual preparation, tireless searching, and extraordinary states of mental illumination will bring them to light. "Behold the herbs!" Paracelsus says. "Their virtues are invisible and yet they can be detected. . . . nothing is so hidden in them

---

[12] Ward and Wilkins, *Vindiciae academiarum*, p. 46.
[13] See, e.g., Webster, *Great Instauration*, pp. 274ff.

that man cannot learn of it. . . . It is God's will that nothing remain unknown to man as he walks in the light of nature; for all things exist for the sake of man. And since they have been created for his sake, and since it is he who needs them, he must explore everything that lies in nature."[14] This reading of inner essences is conceived as a form of interpretation; as the configurations of the stars spell out a message that must be read correctly for the physician to learn the outcome of an illness, so the superficial forms of plants are signs of their inner powers. As voices, gestures, and faces reveal the underlying character of a person, so plants too have their physiognomy and their speech.[15]

This semiotic conception of nature helps to explain the much-remarked obscurity of Paracelsian writing, its gestural quality, its apparent curtailment of the didactic intention. Where teaching and the ability of the written word to inform leave off, there the experience and searching of the reader begin. Nature itself, the wisest of teachers, provides the text in the patterns of the stars, the shapes of plants, the flight of birds, the swarming of bees, the changes of color in the chemist's glass; the role of the physician is only to give hints that may help others read these texts for themselves. To write or depict in a way that is transparently instructional is to usurp nature's role; the true sciences cannot be written down without becoming a dead letter.[16] The things we need to know are disguised; that is one way in which God displays his power. Yet they are not wholly impenetrable mysteries; they require only a careful, sensitive, spiritually informed reading of the book of nature that will grasp the connection between external differentiation and internal virtue.

Experience, in Paracelsus's epistemology, refers not to simple sensory familiarity but to the cognitive identification with the object of knowledge achieved through insight; the learning of the physician is described as a faint "echo," an "overhearing" of the *scientia* that teaches scammony how to purge or pear trees how to grow pears.[17] Nature is instructed by God and is one with the Holy Ghost: "Each day Nature shines as a light from the Holy Ghost and learns from him, and thus this light reaches man, as in a dream."[18] This is the Protestant theme of inner self-illumination turned to the production of results. Mediators and interpreters are useless: the novice can only search and pray, pray

[14] Paracelsus, *Liber de nymphis, sylphis,* in Jacobi, *Paracelsus,* p. 109.

[15] Cf. Jakob Boehme: "The entire visible world and all her creatures is a picture or image of the inner spiritual world: everything which is interior and active shows itself in external markings." *De signatura rerum,* p. 350.

[16] See Temkin, "Elusiveness of Paracelsus."

[17] Pagel, *Paracelsus,* p. 50.

[18] Paracelsus, *De fundamento scientarum sapientaeque,* in Jacobi, *Paracelsus,* p. 181.

and search. The opacity of objects to vision and rationality means that knowledge of occult powers can only be sought in ways that are themselves occult.

Henry Cornelius Agrippa and the elder Helmont teach that the discursive, sequential presentations of school philosophy cannot reach to the inner essences of things. "Every kinde of true, or intellectual knowledge," Helmont states, "is not to be demonstrated."[19] Discourse and "reason" serve only to retake knowledge; they cannot lead to "things not had," which can only be acquired by "invention or gift." Agrippa describes the acquisition of knowledge in his *De occulta philosophia* as "non successione, non tempore, sed subitaneo momente quod cupit assequitur"—the mind grasping in an instant what it seeks[20]—and Helmont speaks of the understanding, which "partakes of an unlimited light, is perfected without weariness and labor."[21]

This rejection of discursive knowledge is matched by an aspiration to the angelic state. The soul, Agrippa says, when freed of disturbing influences, "needs neither memory nor instruction, but by her understanding emulates the angels."[22] Our understanding, Saint Thomas noted, which proceeds by combining or distinguishing concepts, by reasoning from premises to conclusions, is due to the "dimness of the intellectual light in our souls." The angel is a "pure and brilliant mirror who sees manifold things in a simple way and changing things in an unchanging way." As Augustine and pseudo-Dionysius knew, the angels gather their knowledge not from "scattered discourses," but by an immediate intuition; their intelligence is "alight with the penetrating simplicity of divine concepts."[23]

We can detect a common theme in the antilogical empiricism of occult philosophies: the scholastic restriction of human knowledge to what can be achieved by the "combining and distinguishing of concepts" ensures its barrenness. The nondiscursive knowledge of the angels can be won by human beings who seek by another route, whether this is through Paracelsus's "overhearing," a reading of divine visual language, Helmont's inspired trance states, or Agrippa's and Ficino's analogical exploration of macrocosm-microcosm relations. "Every creature," says Oswald Croll, "in this ample Machine of the World, in which the Invisible Creator exhibits himself to us to be seen, heard, tasted, smelt, and handled, is nothing else but the shadow of God."[24]

[19] Helmont, *Oriatrike*, p. 40.

[20] Agrippa, *De occulta philosophia*, book 3, chap. 53.

[21] Helmont, *Oriatrike*, p. 12.

[22] Agrippa, *De occulta philosophia*, book 3, chap. 53.

[23] Aquinas, "How an Angel's Mind Functions," in *Summa theologica* 14:151.

[24] Croll, *Basilica chymica*, preface.

Occult properties are "footsteps of the invisible God in the Creatures, the shadows and Image of the Creator impresst in the Creatures."[25] And when understanding is achieved, Helmont says, "the things themselves, seem to talk to us without words, and the understanding pierceth them being shut up, no otherwise than if they were dissected and laid open."[26]

Are these reveries or are they not? Bacon says that instead of giving out a "dream of our own devising for the pattern of the world," we must seek "the true vision of the footsteps of the Creator imprinted on his Creatures." The language of footsteps and signatures remains, with a new caution about arbitrariness. Bacon agrees that the problem with received knowledge is its superficiality. "The discoveries which have hitherto been made in the sciences are such as lie close to vulgar notions, scarcely beneath the surface."[27] But note the virtues associated with success in occult research: not patience and modesty, but glorification; not detachment, but a heightened state of self-conscious awareness; not a concern with the visible details of things, but a concern with the reality behind their outer shells and husks. The release of a scent from the interior of a body, the understanding of a spoken command: these are the models for discovery, conceived always in revelatory terms. The scientific revolution, in the sense in which I am concerned with it here, articulates a different response to the problem posed by occult qualities, in which the positive drive to uncover what is hidden and the critique of philosophy remain the same in the absence of this particular and problematic notion of experiential immediacy.

The suggestion of Wilkins and Ward in their essay is that Bacon had correctly understood the need for a new approach to the study of nature, one in which the notions of experience and interpretation would continue to play the dominant role, yet one free of idealism and mysticism. Some apparent sources of meaning are spurious, and in achieving knowledge of the inner and hidden powers of objects, it is not fusion that is required but separation, a proper distance between object and observer. The nature philosophers misunderstood this, as though the cure for the abstraction and remoteness of scholasticism were mystical union, direct teaching by God or the Holy Spirit. They did not understand that the obstacle to knowledge of nature is not just the avoidance of experience, but subjectivity and projection.

The "Interpretation of Nature" is a term employed by Bacon in the context of his attempts to renegotiate the relationship between the in-

---

[25] Ibid.
[26] Helmont, *Oriatrike*, p. 26.
[27] Bacon, *New Organon* 1:vii, in *Works* 8:71.

vestigator and the natural world. He works toward this goal conspicu-
ously in the *New Organon* by a process of triangulation, setting the
philosophy of the schools to one side and the practices of "empirics" to
the other. But his suspicion of the imagination and the influence of
subjectivity is carried over from his earlier attacks on prognostication
and astrology. His chief criticism of logic, like that of everyone else, is
the unfruitfulness of its results. The discursive philosophizing of the
Aristotelians is sterile and fruitless, producing nothing but endless talk.
Its practitioners have shut themselves up

> in the cells of a few authors (chiefly Aristotle their dictator) as their persons
> were shut up in the cells of monasteries and colleges [and] did out of no great
> quantity of matter and infinite agitation of wit spin out unto us those la-
> borious webs of learning which are extant in their books. For the mind of
> man, if it work upon matter, which is the contemplation of the creatures of
> God, worketh according to the stuff and is limited thereby, but if it work
> upon itself as the spider worketh his web, then it is endless.[28]

Monastic philosophy, by contrast with occult philosophy, is dull and
empty but not projective. Elsewhere, philosophizing is perverted by its
existential import. It is the spontaneous tendency of the mind to shape
the world according to its own prejudices, habits of association, and
desires that vitiates the content of the false sciences. The "tincture of the
will and affections" imposes on the world an imaginary system of signs
and correspondences that the mind then pretends to discover and de-
code. The false sciences are like poetry, which "doth raise and direct the
mind by submitting the shows of things to the desires of the mind," by
contrast with reason, which "doth buckle and bow the mind unto the
nature of things."[29] The magicians and Paracelsians are shut up in their
systems in which they are guided by similitudes, types, and parables,
which they seek out endlessly in macrocosmic-microcosmic relations.
But the future is not encoded in the flights of birds and the swarming of
bees or in the patterns of the stars; the inner powers of a thing are not
written on its surface. The generation of pseudocontent by the mind
then takes two forms: empty discursiveness, quarrels, and disputes
about terms that correspond to nothing in nature and whose meaning
cannot be fixed; and ideology, the desire to remake the world in an
image favorable to expectations and interests. The *idola* of nature and
culture are either inborn or so inbred that to depose them one can only
disengage oneself from the historical and linguistic community and, by
processes of extreme artificiality, create a context in which they cannot

---

[28] Bacon, *Advancement of Learning*, in *Works* 6:122.
[29] Ibid., p. 203. See, on the distinction, Vickers, "Analogy vs. Identity."

operate. The senses are not to be the direct court of appeal. "For the sense itself is a thing infirm and erring; neither can instruments for enlarging or sharpening the senses do much; but all the truer kind of interpretation of nature is effected by instances and experiments, fit and appointed, wherein the sense decides touching the experiment only, and the experiment touching the point in nature and the thing itself."[30]

This is the function of Bacon's much-maligned method. Insofar as the subtlety of nature is greater than that of the understanding, insofar as sight cannot "penetrate into the inner and further recesses of nature," the scientist must turn away from both vision and reason. The inductive method is the necessary device for dehumanizing the inquiry; it is a "machine" to work upon the stuff of nature, interposed between the mind and its intended subject, which guarantees the objectivity of the results of investigation. It is "artificial, chaste, severe." The aim of the method is not the discovery of laws of the form *All A are B*, but the uncovering of latent structure and latent process. By investigating the phenomena of heat, light, weight, and redness, we shall discover their "forms" not in the metaphysical sense of the Aristotelians but in the sense of their substructure. Heat, for example, is a motion in the smallest parts of a body that renders them "perpetually quivering, striving, and struggling, and irritated by repercussion, whence springs the fury of fire and heat."[31] And this result is embedded within a broader theory of material subtlety, summarized in the conclusion that "[e]very natural action depends on things infinitely small, or at least too small to strike the sense."[32]

The determination of forms or inner subvisible structures should be achieved by a procedure that does not require us to see into, or achieve immediate insight into, the interiors of things. The mind left to itself can provide only "fancies and guesses" as to the form of heat; what is needed is the analogue of the chemist's process of solution and separation, a process that circumvents the limitations of native human perceptual and cognitive faculties. At the end, Bacon says, developing the metaphor, "there will remain at bottom, all light opinions vanishing into smoke, a Form affirmative, solid, true and well defined."[33] To determine the form of heat we begin by compiling a "presentation to the intellect" of all positive instances of heat: the sun's rays, volcanoes, lightning, flame, burning solids, wool, bodies rubbed violently, damp hay, quicklime mixed with water. Next follow the negative instances: the rays of

[30] Bacon, *New Organon* 1:l, in *Works* 8:82.
[31] Bacon, *New Organon* 2:xx, in *Works* 8:215.
[32] Bacon, *New Organon* 2:vi, in *Works* 8:174.
[33] Bacon, *New Organon* 2:xvi, in *Works* 8:205.

the moon, fibrous substances such as linen, quicklime sprinkled with oil. The third table of degrees or comparisons of heat lists factors that increase or decrease the heat present: motion and exercise in animals, wine, fever, the position of the sun and planets, wind.

The reaction of the modern reader is that this is its own form of delusion: Bacon has been carried away by his own chemical metaphor; he has tried to operate purely on the level of presentations, using his tables to move from the immediate phenomena of nature to a knowledge of inner structure in a way that cannot produce results. For his systems of classification are themselves the products of superficial perception and reflection and theoretically so unfounded (he is unable to distinguish between the heat of pepper, the heat of oxidation, and the heat of friction) that his method cannot be the route to the knowledge of a kinetic theory of heat, which is in the end only a sort of guess. Perhaps it is a good guess, based as it is on the observation that a smaller body is heated faster than a larger one, which tends to show, as he says, that heat is "in some sort opposed to tangible matter."[34] But whether the method of tables is the embryonic form of modern statistical reasoning or was replaced by it is again a question of perspective: what concerns me here is that Bacon had proposed, for the first time, a route to the knowledge of the occult, specific properties and virtues of things that proceeded discursively and disinterestedly. The work of searching out forms demands application and exactness, not sensitivity or spirituality, and the investigator's exchanges with the subject matter are not made easy, pleasant, or existentially meaningful. The teasing out of the form of heat generalizes and weakens the investigator's particular interest in feathers, the sun, lime, and life.

But there is another value present here besides that of disciplined detachment: the ideal of pure and undistorted reception. Here it is not a question of keeping the imagination in check, compensating for its overactivity by procedures of exaggerated artificiality, but only of returning to an originally uncorrupted capacity for undistorted reception, what Bacon calls "that commerce between the mind of man and the nature of things" supposedly lost or perverted in the Fall. "All depends," he says, "on keeping the eye steadily fixed upon the facts of nature and so receiving their images simply as they are."[35] Elsewhere he speaks of himself as "dwelling purely and constantly among the facts of nature," withdrawing from them "no further than may suffice to let the images and rays of natural objects meet in a point."[36] Ideal optics is contrasted

[34] Bacon, *New Organon* 2:xiii, in *Works* 8:203.
[35] Bacon, *The Great Instauration* and *Plan of the Work*, in *Works* 8:53.
[36] Ibid., p. 34.

with the deceptive: the individual betrayed by idols is like a "false mirror" that "refracts and discolors the light of nature."[37]

Bacon's attitude toward the usefulness of instruments for sensory enhancement is, however, guarded. Well known is his remark that the greatest hindrance to the human understanding is constituted by the "dullness, incompetency and deceptions of the senses."[38] But despite his reference to focal length, which implies some familiarity with the theory of the lens, there is no evidence that Bacon was able to regard artificial lenses as especially useful devices to supplement the natural optical system of the human eye. True, Solomon's house, the scientific country resort described in the *New Atlantis* of 1627, has a full stock of optical apparatuses: telescopes, eyeglasses, and microscopes, "Helps for Sight far above Spectacles and Glasses in Use." The guide describes to the facility's visitors "Glasses and Meanes to see the small and Minute Bodies, perfectly and distinctly: As the Shapes and Colours of Small Flies and Wormes, Graines and Flawes in Gemmes which cannot otherwise be seen, Observations in Urine and Bloud not otherwise to be seen."[39] Yet from optical instruments relatively little is to be expected, by comparison with instruments for the mind. Bacon does not seriously conceive of an apparatus that will visually take him down to the constituent substructure of bodies: the inductive method must compensate for the failings of nature and technology as well.

Bacon's early arguments for atomism are arguments for the subtlety of nature and depend in no way on enhanced optics. "There is in things a much more subtle distribution and comminution than falls under view," he states in the *Thoughts on the Nature of Things* (1605), adding that division cannot proceed to the infinite.[40] The phenomena of diffusion and transmission indicate how far it does proceed. A small amount of saffron will tinge a barrel of water yellow, and a correspondingly small quantity of civet will infuse a large set of rooms with its pungent odor.[41] Elsewhere he points out that light and color are able to pass through the solid substances glass and water, indicating the existence of minute passageways through apparently continuous matter.[42] Only the elements of matter cannot be infinitely small or the process of diffusion would be infinite as well. Despite his belief in subvisible particles, Bacon was not, or did not remain, an atomist. Some of his later writings suggest that he was undecided about the existence of the void,

---

[37] Bacon, *New Organon* 1:xli, in *Works* 8:77.
[38] Bacon, *New Organon* 1:l, in *Works* 8:82.
[39] Bacon, *New Atlantis*, in *Works* 8:466.
[40] Bacon, *On Principles and Origins*, in *Works* 10:287.
[41] Ibid., pp. 10:287–88.
[42] Bacon, *New Organon*, in *Works* 8:289.

and the *History of the Dense and Rare* positively excludes it. In the *New Organon* he suggests that the same quantity of matter is capable of folding and unfolding itself in a given space, and even in his early *Thoughts on the Nature of Things* Democritus shares the fate of the rest of Bacon's predecessors in being accused of uncontrolled theorizing. Nor does Bacon rule out the existence of things that do not operate by contact and impulse: "Everything tangible that we are acquainted with," he says, "contains an invisible and intangible spirit, which it wraps and clothes as with a garment."[43] This spirit is responsible for softening, melting, giving shape, producing limbs, assimilation, digestion, ejection, and organization, as well as rusting and other "putrefactions."

Nevertheless, Bacon's rejection of scholastic forms and qualities is a direct consequence of his belief in material *subvisibilia*, and his arguments anticipate those of Boyle. Glass and water, which are both transparent, may be turned white, glass by being pounded and water by being agitated into a froth. "We find however that nothing has been added except the breaking up of the glass and water into small parts and the introduction of air."[44] Mashing a pear causes it to acquire a sweet flavor, and gems may be made to lose their color by pounding them. Nothing has changed in either case, except that the parts of the body are now in a new position with respect to one another.[45]

It is important to see that corpuscularian theory entered early modern natural philosophy by degrees, helped after 1473 by the circulation of Lucretius's poem *On the Nature of Things* and the account given of Democritus's beliefs by Diogenes Laertius, which produced a reconsideration of the continuist and antireductive views of Aristotle. There was no revolution in which qualities were suddenly relativized to human observers. The restriction of the properties of fundamental entities to magnitude, figure, and motion, the "primary" qualities of bodies, in contrast with the "secondary" properties, which do not pertain to things in themselves but only as they interact with our sensory apparatus, is not simultaneous with the reintroduction of subvisible corpuscles. Though Descartes claims, when speaking as a methodologist and metaphysician, that matter is only extension, his three grades of corpuscles, the luminous, the luminiferous, and the terrestrial, are still qualitatively differentiated when he comes to speak as a physicist. Though Boyle claims that magnitude, figure, and motion are the universal and catholic properties of matter, he does not deny the existence of quasi-

[43] Ibid., p. 275.
[44] Ibid., p. 221.
[45] Bacon, *Thoughts on the Nature of Things*, in *Works* 10:300.

spiritual effluvia; and although Gassendi revived the atomism of Epicurus, he did not attempt to reduce chemical reactions to mechanical
ones. At best we can say that the scholastic *minima naturalia*, the smallest possible particles of bone, milk, wine, and so forth that retain their
full substantial identity, are gradually replaced by minima with fewer
and fewer qualities. Bacon's particles are, he says, not like sparks of
fire, nor drops of water, nor bubbles of air, nor grains of dust; they are
not heavy or light, hot or cold, dense or rare, hard or soft, "such as
those qualities appear in greater bodies; since these and others of the
kind are the results of composition and combination."[46] They do, however, possess a whole catalogue of qualities: matter, form, dimension,
place, resistance, appetite, motion, and emanations. They do not just
jostle against each other, or collide and rebound, but experience nine-
teen different kinds of motion, including motion of resistance, motion
of connection, motion of liberty, motion of continuity, magnetic motion, motion of flight, royal or political motion ("by which the com-
manding parts in any body curb, tame, subdue, and regulate the other
parts"), motion of rotation, and even aversion to motion ("by which
the earth stands still in its mass").[47] By later standards, this is all only
semireductive: animistic and anthropomorphic language remains, as Bacon's Latinate prose shows:

> if we would study nature scientifically, we must find the way to simpler phe-
> nomena. For the principles, fountains, causes, and forms of motion, that is,
> the appetites and passions of every kind of matter, are the proper objects of
> philosophy; and therewithal the impressions or impulses of motions, the re-
> straints and reluctations, the passages and obstructions, the alternations and
> mixtures, the circuits and series: in a word, the universal process of motions.[48]

By contrast, the corpuscularian philosophy, in the classic formulation
given of it by Boyle, gives a univocal sense to motion; motion means
local motion, motion of translation. The same was true earlier for Descartes. But there is a difference. For Descartes, solidity implies rest:
"Fluids are bodies made up of numerous tiny particles which are agitated by a variety of mutually distinct motions; while hard bodies are
those whose particles are all at rest relative to each other."[49] For Boyle,
intestine motion is occurring always, in solids as well as liquids. This is
easily proved by experiment for the case of liquids; when aqua fortis
and silver are mixed in water and a copper rod is inserted, the particles
of silver will swim to the copper and be deposited. But there are also

---

[46] Ibid., p. 347.
[47] Bacon, *New Organon*, in *Works* 8:302ff.
[48] Bacon, *Thoughts on the Nature of Things*, in *Works* 10:296–97.
[49] Descartes, *Principles of Philosophy* 2:54, in *Philosophical Writings* 1:245.

examples of change in a solid substance that has been left to itself that indicate an internal rearrangement of its parts. A compressed spring loses its springiness, wood becomes seasoned, magnetic power is acquired and lost, spots move in a piece of turquoise, and so on.[50]

Anyone could have noticed these phenomena and made the same deduction Boyle did. And Bacon himself produced no new arguments for the existence of material, quality-reduced *subvisibilia* that had not been adduced or could not have been adduced by the ancient atomists. Why, then, did seventeenth-century theorists feel confident in making the assertion that visible phenomena depend on the subvisible texture and subvisible motions of subvisible entities? Why did they begin to think consistently in corpuscularian terms? Descartes gives in the *Principles of Philosophy* a curious a priori proof of the subtlety of nature. Motion, he says, would otherwise be impossible. If a body moves from E to G it must successively fill all spaces between E and G, but "this is impossible unless part of that matter adjusts its shape to the innumerable different volumes of those spaces."[51] A moving body must thus displace vast numbers of particles occupying arbitrarily small volumes. If the world is a plenum—and, for Descartes, any other hypothesis was inconsistent with the transmission of light from the distant stars—everything, including what the naive call empty space, must be made up of particles that are infinitely, or indefinitely (he does not make a distinction here), divided.

But one is justified in wondering how convincing arguments this abstract, this assumption ridden, in a word this *scholastic*, were to anyone, given that they could be mobilized on all sides. The problem of assuming matter to be both real and either infinitely divisible or arbitrarily divisible just so far pushed foundational thinkers who cared deeply about intelligibility to a rejection of material atomism. Like Aristotle, Leibniz appreciated that the continuum could not be resolved into discrete points or generated out of them. But one might, like Sennert and Gassendi, dismiss the mathematical problem of the composition of the continuum as irrelevant to the question of physical atomism.[52] Pure reason could not decide the issue either way, and it was only in the context of promised applications that arguments about invisible substructure could engage a large number of followers. Even Leibniz recognized a certain pragmatic value to the corpuscularian hypothesis, despite what he regarded as its incoherent foundations. If Descartes had not presented his philosophy as one that would further the realization

[50] Boyle, *History of Firmness and Fluidity*, in *Works* (1772 ed.), 1:446ff.
[51] Descartes, *Principles of Philosophy* 2:35, in *Philosophical Writings* 1:239.
[52] See Joy, *Gassendi the Atomist*, esp. pp. 103ff.; Meinel, "Das letzte Blatt."

of practical ends, particularly in medicine, it is doubtful that many would have been convinced by his reasoning against the plenum. The modern revival of corpuscularianism was successful because reasonable arguments against it could be shelved in view of its technological promise. Corpuscularianism might supply, in the usual philosophical sense, the metaphysical foundation of a technological program. But it was the desire to implement the program and faith that it could be implemented that determined its positive reception.

It has been pointed out in this connection by Keith Hutchison that the advantage of the corpuscularian hypothesis was that its reconception of the scholastic distinction between occult and manifest qualities permitted occult qualities to be rationalized and, in principle, controlled. Following Hutchison's account for the moment, one can observe that the scholastics distinguished between the manifest qualities—the color, shape, consistency, taste, and smell—of a substance, which permit us to recognize it as one of its type, and the occult qualities of a substance, which are nonsensible and not directly inferable from manifest qualities. "[O]ccult or hidden Qualities are those, which are not immediately known to the Sences, but their force is perceived mediately by the Effect, but their power of acting is unknown. . . . We perceive the Actions but not the qualities whereby they are effected."[53] Where the powers of natural things are concerned, the power of the sun to blanch, of rhubarb to purge and arsenic to poison, of the egg to become a chicken, it is evident to everyone that there is something invisible and even "interior" that explains this, for ordinary sensory familiarity does not reveal any such tendencies. Innocuous appearances—the indifferent powderiness of poison and talc—disguise real differences.

The attribution of an occult quality to a given substance is purely a posteriori, following repeated observation of its effects. The power that produces them is outside the scope of legitimate science; magnetism, Aquinas states, is beyond human comprehension.[54] This position naturally implied that the discovery and exploitation of at least some occult qualities could only be the province of magicians assisted by demons.[55] Trial and error constitutes the only permissible route to knowledge in the scholastic scheme: there are no shortcuts that are not dangerous and morally evil, and pharmacological innovation is no simple matter. One should not suppose, however, that there was any consciousness here of an impediment. Given the long experience of humanity, especially with regard to the curative arts, it was hardly to be supposed that further

[53] Daniel Sennert, *Thirteen Books of Natural Philosophy*, in Hutchison, "What Happened to Occult Qualities," p. 234.

[54] Hutchison, "What Happened to Occult Qualities," p. 238.

[55] Ibid., p. 236.

trial and error would produce much new knowledge. People in most periods and in most conditions of life have felt that they knew and understood a great deal about the world. Only magic and science project before them a horizon of things someday to be known; neither common sense nor philosophy does this. The Paracelsian belief that every disease is curable by some as yet undiscovered drug was anomalous in this respect.

To develop some of the implications of Hutchison's analysis, one might observe that the acceptance of a division into manifest and occult, and the belief that there were effects that could be experienced even if their mode of production itself could not be known, is one manifestation of the stasis associated with scholastic philosophy. The verbal, disputatious, inconclusive character of scholastic natural philosophy, which its opponents attacked in the harshest of terms—mocking especially what seemed by their standards the tautological, uninformative character of explanations phrased in terms of virtues and faculties: fossilizing, dormitive, expulsive, rubefacient—was not entirely the fault of an ascetic otherworldliness, but a sensible reduction of inquiry to topics and questions that could be handled within the framework available.

From the point of view of positive science, some occult qualities of the Renaissance corresponded to actual characteristics of substances, while others proved to be the object of purely superstitious beliefs. The lodestone does have the power to draw iron, rhubarb the power to purge, and the sun to bleach, but other powers that had been well attested to by the ancients, such as that of the remora fish to stop a ship, have proved not to exist; to borrow an example of Bacon's, a salamander burned on top of a house cannot raise a thunderstorm. As a result, occult qualities, as Hutchison argues, are not, in the first instance, qualities deemed not to exist by corpuscularian reformers. On the contrary, some of the appeal of the corpuscularian philosophy lay precisely in its ability to establish connections a skeptic might have doubted. Powers such as that of the weapon salve to cure wounds when the weapon that made them was anointed with it were reinterpreted as the effects of the action of small invisible particles or expelled or escaped effluvia. Because these phenomena, having been explained, were no longer occult, the result, according to Hutchison, was that the word "occult" ceased to refer to qualities that were known to exist but could not be explained and came to be a term for the scientifically unintelligible, for what cannot be explained by recourse to invisible particles.[56] It remains in philosophical discourse as a pejorative term, suggesting, as in the celebrated case of the Leibniz-Newton dispute about gravity, that

[56] Ibid., p. 233.

the person referring to the quality does not really know what he or she is talking about.

Finally, Hutchison argues, even the so-called manifest qualities were revealed not to be real qualities of objects, but the effects of the same subvisible but material particles that produced the so-called occult qualities by acting directly on the sensory organs, nerves, and brain. In his somewhat paradoxical formulation, all qualities turn out to be occult in the Aristotelian sense.[57] But one might just as well say that all qualities become manifest in the sense that each is assumed to be associated with some potentially visualizable arrangement of corpuscles that enter into some visualizable interaction with our sensory apparatus.

To say as Hutchison does that the scientific revolution involved a reframing of the problem of occult qualities is to say something profoundly right. The reduction of both manifest qualities and occult qualities to effects of the primary properties of bodies in the treatises of the corpuscularians offered a new way of construing the inaccessibility of the latter, which is no longer an inaccessibility in principle but only in practice. "Not being able," as Charleton says of the Aristotelians, "ever to explicate any Insensible Propriety [sic], from those narrow and barren Principles: they thought it a sufficient Salvo for their Ignorance, simply to affirme all such Proprieties to be *Occult*; and . . . they blushed not to charge Nature Herself with too much Closeness and Obscurity, in that point, as if she intended that all Qualities, that are *Insensible*, should also be *Inexplicable*."[58] Statements of this rejection of powers and virtues as irremediably occult are frequent. As Descartes says, "there are no qualities which are so occult, no effects of sympathy or antipathy so marvelous or strange, nor any other thing so rare in nature . . . that its reason cannot be given by [the principles of mechanical philosophy]."[59]

Yet one may doubt that the demonstration of the intelligibility of occult properties, their accessibility in principle through hypothesized schemes involving invisible effluvia composed of subvisible particles, would have sufficed to establish the corpuscularian scheme. Conversely, it is unclear that an increase in operative knowledge could have been thought to follow from the mere representation to the intellect of such a scheme. Certainly that was the Baconian promise; knowledge of forms *is* power. But this promise would have seemed empty if it had not been believed that help for the senses was at hand, help that would not only confirm the truth of the corpuscularian hypothesis, but actually reveal

[57] Ibid., p. 244.

[58] Charleton, *Physiologia Epicuro-Gassendo-Charltoniana*, chap. 15, "Occult Qualities Made Manifest," p. 342.

[59] Descartes, *Principles of Philosophy* 4:187, in *Oeuvres* 9:309.

directly to the senses, and so render manipulable, the inner constitution of things on which their active powers depended. Lost on the highways and byways of his fabulous inductive method, Bacon did not complete his thought.

Evidence that theoretical corpuscularianism gained strength from new visual experiences is to be found in a seeming inconsistency: the corpuscularians sometimes attack Aristotle and the scholastics as overdependent on their senses and so as "superficial," while at the same time their main arguments for atomism are drawn from sensory experience. Their position is rendered consistent if one remembers that they were impressed by a fact Aristotle did not know: the magnifying lens shows that bodies under magnification have another texture from that which presents itself to the naked eye. This experience enabled them to project, as Aristotle could not, another material world beneath the world we see. Confronted with such microscopical experiences, an Aristotelian would have had either to revise the ontology of qualities and theory of qualitative change and its causes, or else to regard the lens as an illusion-producing device.

Some proponents of corpuscularian theory seemed to place the level where effects were produced permanently out of visual reach. Gassendi echoes Bacon in saying that "the cleverness of nature begins where clever and sharp-sighted perception of nature ends," expressing the view that the true constituents of objects are just too small ever to be seen by mortal eyes.[60] But appearances under the microscope could be taken as evidence for atomism. Charleton remarked in 1654 that the ancients who lacked optical instruments might have been skeptical about the existence of atoms, "but for us, who enjoy the advantages thereof, and may, as often as the sun shines out, behold the most laevigated Granule of dissolved Pearl," the situation is different. The irregularity of objects seen under the microscope is an argument against the Aristotelian theory of homogeneous substance. As Charleton says, "The superfice of no body can be so exactly smooth and polite, as to be devoid of all unevenness or asperity, every common Microscope discovering numerous inequalities in the surface of even the best cut Diamonds, and the finest Chrystal."[61] Spiders and flies may run up and down on Venice glass, which they could not do if there were not cavities for the hooks of their feet to grip, and glass would not fog up if there were no valleys in which the moisture could settle.[62] All this shows, on Charleton's interpretation, that the perceived continuity of

[60] Gassendi, *Opera omnia* 3:16.
[61] Charleton, *Physiologia Epicuro-Gassendo-Charltoniana*, p. 97.
[62] Ibid., p. 154.

substances is due to a perceptual blurring of the little mountains and valleys on a surface. Water must be like fine sand, though even the microscope cannot see its particles; the uniformity of a piece of metal is as illusory as the uniformity of a sheet of paper or a piece of woven cloth, which do reveal their irregularity under the lens. All this showed, Charleton thought, that "no one need fear the section of an Atome." This argument led, as Christoph Meinel points out, to an antinomy of the microscope: vision is trustworthy and truth-revealing; vision is superficial and deceptive.[63] If the microscope had revealed the perfect smoothness of ordinary substances, then there could not be atoms, for substances would then be continuous just as Aristotle had said. But were the surfaces of atoms themselves irregular or smooth? The microscope seemed to reveal only hidden roughness. But if no surface is really smooth, atoms cannot have an assignable shape or size; they are not determinate and primitive beings. Charleton did not think this far, or this philosophically, ahead; he believed himself simply to have direct evidence against Aristotle.

Other examples of the role of experiences with lenses in creating a favorable context of reception for corpuscularianism are scattered through the literature. In the *Zootomia democritaea* of Marco Aurelio Severino, published in Nuremberg in 1645,[64] the author argues that a "resolutio in indivisibilia" or a "resolutio ad minimum" through a microanatomy of vegetables, insects, other animals, and humans should take the place of ordinary dissection. Anatomy should reach the fundamental units of living things, but this is only possible when it becomes "artificiosa." Some years later, Leibniz found in Hooke's *Micrographia* the resources to reinterpret Aristotle in a contemporary mode (that is to say, to reconstruct him in unrecognizable form) by reinterpreting the Aristotelian varieties of motion—generation, corruption, increase, decrease, and alteration—as the results of local motion in the microworld. "I observe in advance," he says,

> that numerically the same change may be the generation of one being and the alteration of another; for example, since we know that putrefaction consists in little worms invisible to the naked eye, any putrid infection is an alteration of man, a generation of the worm. Hooke shows similarly in his *Micrographia* that iron rust is a minute forest which has sprung up; to rust is therefore an alteration of iron but a generation of little bushes. Moreover, generation and corruption, as well as alteration, can be explained by a subtle motion of parts.... This is why foaming water is white, for it consists of innumerable little bubbles, and each bubble is a mirror.... Such considerations make it clear that colors arise solely from a change of figure and posi-

[63] Meinel, "Das letzte Blatt," p. 13.
[64] See Belloni, "De la theorie anatomistico-mechaniste," p. 100.

tion in a surface. If we had space, it would be easy to explain light, heat, and all qualities in the same way. . . . an essence differs from its qualities only in relation to sense.[65]

It is usual to interpret such passages as involving merely the application of a corpuscularian theory to the study of the body and bodies in general, for the conviction expressed here does not depend on the author's belief that he has actually perceived atoms or the minute corpuscles composing all objects; he need only have noted a "texture" radically different from surface texture. But some natural philosophers did hope to see them: "I am apt to suspect," Robert Boyle wrote in 1663,

> that if we were sharp-sighted enough, or had such perfect microscopes, as I fear are more to be wished than hoped for, our promoted sense might discern in the physical surfaces of bodies both a great many latent ruggednesses, and the particular sizes, shapes and situations of the extremely little bodies that cause them and perhaps might perceive . . . how those little protuberances and cavities do interrupt and dilate the light . . . according to the nature and degree of the particular color we attribute to the visible object.[66]

Boyle was in time criticized for his suggestion that the microscope might help us understand what color consists in. Locke argued that the relationship between primary and secondary qualities was such as to preclude this understanding. But both passages tend to confirm the thesis that belief in the distinction between primary and secondary qualities was itself a product of experiences with the magnifying lens, or, to be more precise, the thesis that the reason this old distinction of the ancient atomists came to be seen as compelling, despite the Aristotelian arguments against it, was that the microscope showed that the surface appearance of a thing might be entirely different from its magnified texture. If roughness could appear as smoothness, if tiny, transparent bubbles could appear as white, why should not a certain kind of surface texture appear as—yellow? Leibniz's own repeated arguments that qualities emerge as the result of confused perception of underlying texture, just as a mixture of blue and yellow powder produces the new appearance of green, were simply an attempt to justify this alternative way of considering sensory properties.[67]

Think again of Aristotle's use of ordinary sense experience as an arbiter of disputes. His arguments against atomism, for example, are a combination of a priori arguments against the possibility of indivisibles and a void, and arguments drawn from its inconsistency with perception.

[65] Leibniz, letter to Thomasius, in *Leibniz*, pp. 96–97.

[66] Boyle, *Works* (1744 ed.), 5:680.

[67] See, for example, Leibniz, *New Essays*, in *Sämtliche Schriften und Briefe*, series 6, 4:403.

But this repudiation of the atomists' account of coming to be, alter-ation, and passing away as atomic accumulation, reshuffling, and the dissolution of atomic aggregates left him with what appeared to the moderns to be an essentially verbal solution to the problem of change: change in general is the transformation of power into act; change of quality is the intension or remission of form. The alternative, that col-ors, smells, and tastes can be generated ex nihilo, as the atomist posi-tion seemed to him to imply, was simply out of the question. And for a long time philosophers found this reasoning convincing and the prof-fered analysis of change profound, until they saw that the generation of qualities from invisible atoms need not be considered as ex nihilo. Only at that point did the theory of virtues and powers, and with it the asso-ciated language of form and matter, real qualities, acts, and potentials, suddenly seem—as it never had before—empty.

One should not, then, underestimate the a priori satisfactoriness of the Aristotelian account of qualities, which had endured, after all, through thousands of years of sharp-minded philosophy, or overesti-mate the original a priori satisfactoriness of the corpuscularian hypoth-esis. We need not take the corpuscularians' word for it that the Aris-totelian account had been all along a strangely overlooked tautology whereas atomism and reductionism were both probable and informa-tion laden.[68] Recall in this connection the problem of the gap between primary and secondary qualities: the production of the latter from the former was asserted but not well explained by the corpuscularians. They did not really succeed in making them intelligible. But before Locke, whose problematizing of that relation was part and parcel of his critique of microscopical knowledge, none of the moderns was seriously troubled by this. When the data are organized somewhat differently, the apparent failure of the pre-Lockeans to be worried by the gap between primary and secondary qualities is no longer a puzzle.[69] The tautologous character of Aristotelian explanations came into being only when natu-ral philosophers became convinced that the interiors of things could be revealed to the eye. They were not initially bothered by a gap, just be-cause microscopical experience seemed to give positive examples of the production of new, emergent qualities. It was only during the later phase of the microscope's introduction that its failure to make good on its promises caused skeptical voices like Locke's to be raised.

According to Robert T. Gunther, the dust from the brown patches on the back of mature fern leaves was once regarded as a valuable charm

---

[68] See Hutchison's more recent article, "Dormitive Virtues, Scholastic Qualities."

[69] Antimechanists did not, any more than Aristotle, approve of reductive accounts of qualities, but their opposition does not concern me here.

capable of making a person invisible, for ferns may be found surrounded by their tiny progeny but are without flowers and seeds. The
principle of replication was held to be invisible, though contained in
this brown dust, and to work by a kind of sympathetic magic.[70] As an
example of the logic of the occult, this is exemplary; one only wonders
whether many people really believed it. By showing that objects and
structures of a hitherto unknown sort existed beneath the level of ordinary perception, the microscope worked against the supposition that
there are sympathetic connections and patterns of copying or replication that are unmediated, or mediated by immaterials.[71] It is true that
many philosophers did, as Hutchison points out, simply replace unmediated sympathetic connections by connections mediated by minute
but material particles. But, as I will argue, the conviction that there are
subvisible material causes of the most obscure phenomena drove out
explanations that involved spiritual entities or correspondences. When
the microscope had reached its full development, Ward and Wilkins
told Webster in the course of heaping scorn on his dream of a "vive and
mystic" anatomy, even the mysteries of animal generation would be
revealed. And that is in fact how it went, not only where the question of
the origins of life was concerned, but also with regard to the puzzles of
illness and death.

Meanwhile, microscopists wondered about the causes of the sharp
tastes of vinegar, pepper, and sage, seeking some clue in their microscopical appearance. And here indeed the copying or replication motif
was still present; the explanations they arrived at, like the theory of
embryological preformation, were semireductive: the sharp edges of
cotton fibers that Leeuwenhoek thought interfered with the healing of
wounds, the tiny spiders alleged to produce the pungency of sage, occupied a sort of halfway point at which analogical thinking still dominated. It would be wrong, then, to think that one is dealing here with
exclusive mentalities, rather than with a displacement over the long run,
that a belief in the production of forms and qualities *ab infra* excludes,
in a single mind, a belief in correspondences and influences *de supra*.
The microscopist Grew, better known for his plant studies, continued to
believe that the dominion of the spiritual over the material world mani-
fests itself in coincidences: that the sun and moon influence diseases and
childbirth, that comets are forerunners of plague, that an earthquake
foretold the death of a Flemish queen, that there is a necessary harmony, a projected concurrence, to the natural and moral orders. But the
rejection of the theory of signatures of the early modern period was tied

[70] Gunther, *Early Science in Oxford* 3:209.
[71] According to Miall, in the Middle Ages it was believed, despite the testimony of the
ancients, that ferns spring from the brown dust on the underside of their leaves, and the
first to say this in print was Valerius Cordus (d. 1544). Miall, *Early Naturalists*, p. 28.

to a general rejection of semiotic conceptions of nature. The belief that nature reveals itself through types, replications, copies, and images, which, like the puzzling types and parables of Christianity, are to be recognized and read by the adept, was subjected to attack. As Bacon questioned the Paracelsian idea that the surfaces of natural things are icons and tokens of their inner powers, Hooke argued that fossil shells are not copies formed as a prank of nature, but are generated by processes of microscopic deposition. The interest in similarities—such as roots exactly resembling a man's hand—that surfaced from time to time in the learned journals of the mid–seventeenth century came in for sharp criticism by the new naturalists. Swammerdam poked fun at the naturalist Goedaert, who saw a human face on a caterpillar's back; Leibniz challenged those who saw pictures in mineral veins, and Hooke was ready with the explanation. "Trees, Hills, Houses and other Perspective Representations," such as those seen in ancient rocks, are only the result of the chance accumulation of tiny particles in the juices that have settled in their clefts.[72]

To what extent can this rejection of a world of reduplications, enigmas, patterns, and correspondences really be connected with the revelations of the microscope? Insofar as the production of form becomes a process to be analyzed in terms of cumulative small changes invisible to the naked eye, the answer is clear. The question of origins, of how a thing—insect, shell, root, comet—came to be, becomes one to be answered by appeal to causal processes acting within or on some less-organized material. On the macroscopical level, we are given stories like Descartes's account of the emergence of the universe from a primordial chaos, differentiated and formed by laws of motion, excluding any intentionality, any formative purpose in nature. But the microscope reinforced this pattern of explanation, reducing the gap between order and chaos. If final causes could not be dispensed with entirely by the new physics, their operation was nevertheless understood as less and less particularized. Above all, the microscope takes away the privilege of surface. What the object looks like on the outside is no guide to what it is in the sense of what it can do; the key to its powers is to be found in its inner invisible structure. And in the interior of things there is no resemblance: here is indeed a new world even if we must call in the language of every day—of ropes, fibers, globules, forests, looms, and children's toys—to describe it. That world was come upon accidentally, as a result of an invention by means of which we are enabled to do what humans ordinarily cannot and need not do. The secrets of things were, in this sense, not meant for us to find out at all; we are always, in

[72] Hooke, "Discourse of Earthquakes," in *Posthumous Works*, p. 436.

this new world, intruders. That nature is a system of signs meant for us to read is no longer a tenable view. Nature is not a book.

One of Hooke's more profound ideas is that knowledge of nature is not won by a direct reading, but by a circular passage from observation to memory and generalization and down again to observation:

> [I]t is to *begin* with the Hands and Eyes, and to *proceed* on through the Memory, to be *continued* by the Reason; nor is it to stop there, but to *come about* to the Hands and Eyes again, and so, by a *continual passage round* from one Faculty to another, it is to be maintained in life and strengths, as much as the body of man is by the circulation of the blood through the several parts of the body, the Arms, the Feet, the Lungs, the Heart, and the Head.[73]

One expects, nevertheless, some transfer of metaphor, some sentimental persistence of language. And indeed we find it, in Swammerdam's *Biblia natura* and in the depths and banalities of physico-theology, in the reference to the writing in "Nature's smallest hand" in Cowley's "Ode to the Royal Society," in Hooke's reference to the "signatures" of seeds of time, even in Bacon's reference to science as recovering the "true vision of the footsteps of the Creator imprinted on his creatures." These images of stamping, printing, and reading persisted into the eighteenth century in Bishop Berkeley's visual language and Kant's aesthetics, these repetitions testifying to the depth of the metaphor. But the new notion of reading was a decoding of an arbitrary alphabetical language in which the combination of elements had to be broken apart by analysis; it was not, except when displaced from science to metaphysics and aesthetics, the interpretation of a symbol pointing beyond itself. Nineteenth-century "scientific" physiognomy is reduced to its core meaning—the display of moral and intellectual character in human facial features—and finally loses its empiricist pretensions altogether.

Bacon, in the first quarter of the seventeenth century, used optical metaphors to describe cognition somewhat selectively. Clarity, focus, and freedom from distortion, rather than magnification, were the optical qualities that interested him. It is significant how magnification too became, with the deployment of the microscope, a feature of the prelapsarian excellence of vision. This conceit is present in Joseph Glanvill's "skeptical" work, *The Vanity of Dogmatizing* of 1661, and in its later version, the *Scepsis scientifica* of 1665. Before man's first sin, Glanvill says, Adam's senses were "without any spot or opacity; to liken them to the purest Crystal, were to debase them by the comparison." He not

---

[73] Hooke, *Micrographia*, preface.

only felt the earth's motion, his "natural Opticks" let him see both far and small without "Galileo's tube."

> His sight could inform him whether the Loadstone doth attract by Atomical *Effluviums*; which may gain the more credit by the consideration of what some affirm; that by the help of *Microscopes* they have beheld the subtile streams issuing from the beloved *Minerall*. It may be he saw the motion of the blood and spirits through the transparent skin, as we do the workings of those little industrious *Animals* through a hive of glasse.[74]

The prelapsarian state would have been one in which "the circumference of our Protoplast's senses, should be the same with that of nature's activity."

Although the thrust of Glanvill's argument is superficially negative, his point being that our fallen state prevents us from understanding the world around us, he paints a vivid picture of what comprehensive knowledge would come to: knowledge of celestial phenomena, occult effects, perception and sensation, generation and growth: in short, he lays out a broad research program. The problem of knowledge does not reside in anything but a simple mismatch: "our *senses* being scant and limited, and Natures operations subtil and various; they must needs transcend, and out-run our faculties. They are only Natures grosser wayes of working, which are *sensible*; Her finer threads are out of the reach of our dull *Percipient*. Yea questionless she hath many hidden *Energies*, no wayes imitated in her obvious pieces."[75]

That this lament might be taken simply as an invitation to enlarge the scope of the senses, that pessimism might readily be converted into optimism, is plain, and it is not surprising to find that Glanvill later published in 1668 a defense of the Royal Society in his *Plus ultra*, in which he commends the microscope, along with the telescope, the thermometer, the barometer, the air pump, and so forth, saying that "this *Instrument* hath been exceedingly improved of late, even to the *Magnifying* of Objects a *thousand* times, and many useful Theories have been found and explicated by the *notices* it hath afforded."[76] *Plus ultra* has nothing but enthusiasm for its ability to "look into the *minutes* and *subtilties* of things, to discern the otherwise *invisible Schematisms and Structures*," repeating that the secrets of nature are "not in the *greater* Masses, but in those little *Threds* and *Springs* which are too *subtile* for the *grossness* of our *unhelp'd Senses*."[77]

---

[74] Glanvill, *Vanity of Dogmatizing*, pp. 4–6.
[75] Glanvill, *Scepsis scientifica*, p. 51.
[76] Glanvill, *Plus ultra*, p. 57.
[77] Ibid.

In the meantime, Henry Power and Robert Hooke had appeared on the scene as microscope proponents and publicists. Power's *Experimental Philosophy* of 1664 was organized in part as a challenge to Glanvill's earlier book. After acknowledgments to Bacon, Democritus, and the student of insects Thomas Moffett, the preface refutes the charge of human incompetence. To Glanvill's charge that fallen humanity is helpless to see the causes of things, Power asserts that he does not agree that "the Aged world stands now in need of Spectacles, more than it did in its primitive Strength and Lustre: ... certainly the Constitution of *Adam's* Organs was not diverse from ours, nor different from those of his Fallen Self, so that he could never discern those distant, or minute objects by Natural Vision, as we do by the Artificial advantages of the *Telescope* and *Microscope*."[78] It is not a question of restoring a human intellectual empire but of creating one for the first time. If practical optics could only match the promise of theoretical optics and

> that daring Art could perform what the Theorists in Conical sections demonstrate ... we might hope, ere long, to see the Magnetical Effluviums of the Loadstone, the Solary Atoms of Light (or *globuli aetheri* of the renowned *Des-cartes*), the springy Particles of Air, the constant and tumultuary motion of the Atoms of all fluid Bodies, and those infinite, insensible Corpuscles which daily produce those prodigious (though common) effects amongst us.[79]

Here the idol-defeating power of aids to the mind is taken over, in a way Bacon would not have approved of, by aids to the eye. Borrowing Bacon's metaphor, Power argues that "without some such Mechanical assistance, our best Philosophers will prove but empty Conjecturalists, and their profoundest Speculations herein, but gloss'd outside Fallacies; like our Stage-scenes or Perspectives that shew things inwards when they are but superficial paintings."[80] The "modern Engine," as he calls the microscope, would enable us to see "what the illustrious wits of the Atomical and Corpuscularian Philosophers durst but imagine," freeing us from empty conjectures and philosophers' constructions. As the telescope freed us from futile speculation about the heavenly bodies by bringing what was remote near, so the microscope will demystify occult effects.[81] There were some limits; at least some knowledge seemed to lie beyond the horizon. Power says that some have ranted that they have seen aromatic, magnetic, and electrical effluvia with the microscope but denies that this is possible: even the steam of our breath is invisible, so

---

[78] Power, *Experimental Philosophy*, preface.
[79] Ibid.
[80] Ibid.
[81] Ibid.

that observation of these things, along with "Spiritualities ... will, I fear, prove the last Leaf to be turned over in the Book of Nature."[82]

Hooke too was inspired by Bacon. In the preface to his *Micrographia* of 1665 he begins with a diagnosis of the inadequacy of our mental powers, devoting special attention to the deficiencies of our sense organs relative to those of other creatures, and, what is more important, to the lack of proportion between our natural equipment and what he calls indifferently "the Object." As with Glanvill, this disproportion at first appears simply as an addition to the standard list of varieties of human fallibility, but it is one that allows of remediation. For we can add "*artificial Organs* to the *natural*," which

> has been of late years accomplisht with prodigious benefit to all sorts of useful knowledge, by the invention of Optical Glasses. By the means of *Telescopes* there is nothing so *far distant* but may be represented to our view; and by the help of Microscopes, there is nothing so small, as to escape our inquiry; hence there is a new visible World discovered to the understanding. . . . By this the Earth it self, which lyes so neer us, under our feet, shews quite a new thing to us, and in every *little particle* of its matter, we now behold almost as great a variety of Creatures, as we were able before to reckon up in the whole *Universe* it self.[83]

Hooke's rhetoric is ostensibly directed against the philosophy of discourse and disputation, the subtleties of scholasticism. It employs a catalogue of dichotomies: the science of nature, which is the product of "a *sincere Hand,* and a *faithful Eye,*" by contrast with the one that is a "work of the *Brain* and the *Fancy*"; "*solid Histories, Experiments,* and *Works,*" as opposed to "fine *dreams* of Opinions, and *universal metaphysical natures,* which the luxury of subtil Brains has devis'd."[84] The choice for philosophy is between "wandering far away into *invisible Notions*" and returning into "the same *sensible Paths,* in which it did at first proceed."[85]

It was on this score that the moderns could be certain of their much-contested superiority to the ancients, convinced that they were not simply restoring and repeating but building anew. They were going beyond the essential superficiality of the old natural historians, Aristotle and Pliny, in whom we find, says Hooke, both an inexactness and "a needless insisting upon the outward Shape and Figure, or Beauty," a desire to induce in the reader pleasure, divertissement, admiration, and wonder,

---

[82] Ibid., p. 58.
[83] Hooke, *Micrographia*, preface.
[84] Ibid.
[85] Ibid.

but not knowledge.[86] When Hooke speaks the language of restoration, saying that the "*artificial Instruments* and *methods*" may offer "in some manner, a reparation . . . for the mischiefs, and imperfections, mankind has drawn upon it self, by negligence, and intemperance, and a wilful and superstitious deserting the Prescripts and Rules of Nature, whereby every man, both from a deriv'd corruption, innate and born within him, and from his breeding and converse with men, is very subject to slip into all sorts of errors,"[87] he is not seriously proposing the microscope as a means of secular salvation or release from original sin, but only countering the arguments of religious pessimists who employ skepticism under the guise of piety to depreciate research, and of the genuinely pious who see nature as forbidden territory.[88]

The historiography of science has not treated the five classical microscopists, Swammerdam, Leeuwenhoek, Grew, Hooke, and Malpighi, kindly. Bracegirdle says that many early microscopists seemed more interested in the design, construction, and performance of their microscopes than in their use.[89] Rádl says of Malpighi that "it is his own fault that his anatomy excited no new investigations, no new ideas."[90] "Their results," says Singer, "impressed their contemporaries as deeply as they have some modern historians. But none of these microscopists inspired a school, and we have to turn to the nineteenth century for their true continuators. On this account, the 'classical microscopists' must be accorded a less prominent place than the great interest of their biological observations might suggest."[91] He notes, however, that the failure of microscopy to lead to any continuous developing tradition from its founding is "a remarkable fact and one that awaits adequate explanation."[92]

By 1692, Hooke was already complaining of a reaction against the microscope, of boredom and disenchantment. Bacon's doubts about the worth of mere vision-enhancement had turned out in a sense to be justi-

---

[86] Hooke, "A General Scheme or Idea of the Present State of Natural Philosophy and How Its Defects May be Remedied," in *Posthumous Works*, p. 4.

[87] Hooke, *Micrographia*, preface. Cf. Bacon on reparation through the labor and toil of science, in *Works* 8:53–54.

[88] Thus Nathaniel Culverwell: "There is not enough light in any created lamp to give such a bright displaying of an object. Nor is there enough vigour in any created eye, so as to pierce into the pith and marrow of being, into the depth and secrecy of being." *Discourse*, p. 143.

[89] Bracegirdle, *History of Microtechnique*, p. 8.

[90] Rádl, *Geschichte der biologischen Theorien* 1:166.

[91] Singer, *Short History of Scientific Ideas*, p. 281.

[92] Ibid., p. 172.

fied. In later chapters I will address the reasons for this loss of confidence. But the generally dismissive treatment of the classical microscopists seems unjustified, and Singer's puzzlement over the apparent failure of early microscopy is an indication that the failure here is to some extent another historiographical illusion, along with other gaps, lapses, and supposed instances of regression. For the microscope, in addition to the particular reorientation it gave to the study of anatomy, embryology, pathology, and the problem of ultimate origins of life, had broken the philosophical impasse posed by occult qualities. In revealing layer after layer of articulated structure, the microscope gave solidity and accessibility to what theory delivered up as an atomized or mathematized world. Even for those who were, to a greater or lesser degree, skeptical about the actual powers of the microscope, what Hume was later to call "an intricate machinery or secret structure of parts" was irrevocably proposed as a possible object of experience.

While there was a certain degree of mutual validation exchanged between corpuscularians and microscopists, it would be a mistake to assume that their interests or convictions were or remained entirely consonant. "The real, the true, the mechanical philosophy" of Hooke was not firmly affixed to any tightly defined corpuscularian philosophy or to the statements of universal mechanism of the Cartesians. As Glanvill said, though the "Grand Secretary of Nature, the miraculous *Des-Cartes* have here infinitely out-done all the Philosophers went before him in giving a particular and *Analytical* account of the *Universal Fabrick*: yet he intends his Principles but for *Hypotheses* and never pretends that things are really or necessarily as he hath supposed them."[93] Glanvill raises the objection put to Galileo, that omnipotence and infinite wisdom cannot be confined with the boundaries of the human imagination and its "shallow models":

> we may hopefully expect a considerable *inlargement* of the *History of Nature*, without which our *Hypotheses* are but *Dreams and Romances*, and our *Science* meer *conjecture* and *opinion*. For while we frame *Scheames* of *things* without consulting the *Phaenomena*, we do but build *in the Air*, and *describe an Imaginary World* of our *own making*; that is but little a kin to the *real* one that *God made*. . . . For the advancing *day* of *experimental knowledge* discloseth such *appearances*, as will not lye *even*, in any *model* extant.[94]

This is simply another reflected version of the Baconian caution against giving out a dream of our own devising for a picture of the world, a

---

[93] Glanvill, *Vanity of Dogmatizing*, p. 211.
[94] Glanvill, "Dedication to the Royal Society," in *Scepsis scientifica*.

caution that proves again that atomism was no original paradigm of scientific objectivity.

The substantive criticism leveled by the occult nature philosophers against the teachings of the universities and the medicine of the academies was that they were "external." As Paracelsus said, despite all the exertions of Aristotle and his followers, there remained in the kernel of nature only "dust and a withered flower."[95] The Aristotelian intellect, according to Helmont, "beholdeth things only on the outside."[96] Neither naive sense perception nor disputation and argument could supply a knowledge of nature "in her Root and thingliness." The predictive and explanatory powers that differentiate the nonmathematical sciences from other forms of learning were thus necessarily associated, up to the middle of the seventeenth century, with a set of interpretive procedures allied with suprarational techniques. "We require," says Croll, "a much higher ingenuity and subtile Inquisition, than can be obtained by the sight of eyes only."[97] These criticisms of the philosophy of the schools were echoed, not always deliberately, by Bacon and Hooke, who urged the advancement of knowledge beyond "superficies" and outward forms, to the secret and "occult" workings of things, under which Hooke himself included sympathies and antipathies, influxes and the works of fantasy and imagination.

The microscope in the first century of its use appears to have disappointed many historians. But it is wrong to concentrate too keenly on what the classical microscopists did not find, what they misinterpreted or misunderstood, or their lack of organizational sense. What is important is that the microscope both undermined confidence in the manifest image of the world and, in supplying a glimpse of a latent image, gave sense to the idea of a nonoccult interpretation of nature. It showed a morally and theologically permissible way to find a route to the "pith and marrow of being," so binding knowledge to power in a way that the irredeemably fictional mechanical models, as well as the search for linguistic essences, had been unable to accomplish.

[95] Pagel, *Paracelsus*, p. 52.
[96] Helmont, *Oriatrike*, p. 40.
[97] Croll, *Basilica chymica*, preface.

3

Instruments and Applications

SCIENCE differs from natural philosophy not simply and perhaps not primarily in its attention to quantification, but in its use of instruments both for measurement and for the creation of artificial states and experiences. Measurement and computation are robust practices that link ancient astronomy with every modern scientific branch of inquiry; sundials, astrolabes, and surveying compasses, which, in abstract terms, are devices serving both to orient the observer and to apportion units of time or space, were produced for centuries. Yet as Derek de Solla Price has argued, both the making and the deployment of instruments have certain puzzlingly autonomous features. In ancient and medieval times, he points out, people did not consistently use their instruments for time-keeping, navigation, or surveying, or take full advantage of their capacity for accuracy. Instruments were rather made, exchanged, and kept as models or symbols of the possession of certain kinds of knowledge. Ancient sundials, for example, were not so much used to tell time or regulate the activities of the day as to symbolize the daily movement of the sun; in astronomy, naked-eye observations were preferred to the available apparatuses for measuring degrees of arc.[1] Histories of scientific instruments are accordingly apt to reflect the connoisseur's interest in the introduction of particular design features, in the details of instrument manufacture, and in their beauty as objects of art rather than their actual uses. The historiography of science, by contrast, has frequently assumed that instruments function transparently, that they serve to collect information that is available in nature, and that particular or general features of the apparatus are of no direct relevance to the gathering of new knowledge. What is needed is an examination of the interaction between the history of technology and the history of science that takes into account the problems that arise in connection with the idea that a science based on the use of machines and instruments gives a truer, better, or deeper account of the world.

The material and mechanical bias of the seventeenth-century scientific revolution had already been noted, along with the convergence of craft and theory in an array of previously unimagined devices. The ingenious

---

[1] See Price, "Of Sealing Wax and String" and "Philosophical Mechanism and Mechanical Philosophy."

Hooke produced, besides the microscope associated with his name, the air pump required for Boyle's experiments on the vacuum and the spring of air, a hygroscope for measuring humidity using the beard of a wild oat that uncoiled itself under the influence of moisture, gauges, quadrants, clocks, an arithmetical machine, and a diving bell. Price detects important consequences in this shift from modeling or representation to application, arguing that it was with the introduction of the telescope that instruments took on a new importance in the acquisition of knowledge. Galileo's instrument-mediated perception of the phases of Venus, the satellites of Jupiter, and the mountains on the moon were, Price argues, both necessary and sufficient conditions for the acceptance of the Copernican theory. "Without the telescope no worthy theorist would ever have made the switch to a model of the universe whose planetary kinematics seemed so much more complex and less accurate than the accepted model."[2]

By the middle of the eighteenth century, the world was full of telescopes and microscopes too, "some better, some worse,"[3] and few doubted that physical interactions in the microworld were the basis of most observable changes; I have sought to capture the earliest phase of this realization. But the positivism that supposes that technologies and procedures are capable of driving discovery by themselves needs qualification in many respects. History suggests that new technologies and procedures come to be valued and exploited only in the context of certain hopes and expectations. On the philosophical side, the use of instruments raises a spectrum of difficulties concerning trustworthiness and novelty that still dominate discussions about empiricism, theoretical entities, and scientific realism. We need then to ask how this ideal of instrumentally mediated knowledge succeeded in getting itself established, how the notion of a scientifically revealing experience became uncoupled from the broader notion of experience in general. And we need to ask how confidence was established in experiences that were profoundly unnatural, produced by an awkward and frequently troublesome apparatus that interposed itself between observer and subject matter. We know in fact that the culture has not entirely assimilated its own science and that an undercurrent of mistrust attaches to specialized knowledge at the points at which it invokes the demonic in its implied power over nature, and at which it has had to forsake immediacy and immediate intelligibility.

Had the ancients, one might wonder, really been entirely ignorant of the use of lenses, as Henry Power and Robert Hooke, who called attention

[2] Price, "Of Sealing Wax and String," p. 52.
[3] Gunther, *Early Science in Oxford* 3:289.

to the ways in which the new Christian science was unlike and superior to the old pagan science, liked to imply? Scattered references to feats of miniaturization, especially in jewelry making, can be found in ancient authors; these were once taken as showing either that they knew how to grind lenses or that they were able to take advantage of the refractive properties of naturally occurring crystals. Callicrates was famous for his flies modeled in ivory, Strabo was said to have written the *Iliad* on a single sheet of paper stored inside a nut, and Pliny, who described insects in fine detail, mentions Nero's use of an emerald held close to the eye at gladiatorial games, presumably to help him see at a distance, in his *Natural History*.[4] More-recent historians have tended to regard these stories with skepticism, arguing that extreme nearsightedness rather than the use of magnifying instruments can account for the fineness of detail achieved by ancient craftspeople.[5] At least one convincing reference to the use of magnifying lenses for the study of nature before the seventeenth century exists, however; the medieval Jewish philosopher Gerschom said in the first half of the fourteenth century that it would be possible to look at insects through a flask of water or with a "burning glass" or concave lens, and that the lost knowledge of the ancients regarding them might thereby be restored.[6] But Gerschom proposed this only as a possible use for burning glasses; he did not claim to have performed or to know of observations carried out by these means. Alexander Benedictus, a physician of Verona, is believed to have made observations with a lens around 1508, finding "wormlets" in skin, a kidney, a lung, teeth, and cheese.[7] John Mayall called attention to a portrait of Pope Leo X painted around 1520, in which the pope holds a lens for examining miniatures, suggesting that both painting and art appreciation benefited from the use of single lenses.[8]

By the late twelfth and early thirteenth centuries, the optical theory of the Arabs had been disseminated and studied in Europe through the *Perspectiva* of Ibn-al-Haytham, or Alhazen, who had explained geomet-

---

[4] Pliny, *Natural History* 87.16. See Harting, *Das Mikroscop*, pp. 573–74.

[5] See Mayall, "Cantor Lectures on the Microscope," vol. 34 (1887): 987–89.

[6] "Many of the varieties of animals mentioned by Aristotle are unknown to us, especially the varieties whose members have particularly small bodies. For although we perceive them, it is difficult for us to come to know their forms and characteristics on account of their diminutive size. I have however an idea for an invention by means of which we may come to know the organs of animals which are so tiny that they escape our perception; we should observe them with the help of things which show us the perceived object larger than it is, like, for example, the burning glass." Gersonides, "Commentary on Averroes's Commentary on Aristotle's 'Book of Animals,'" in Freudenthal, "Human Felicity and Astronomy," p. 62.

[7] Singer and Singer, "Development of the Doctrine," p. 190.

[8] Mayall, "Cantor Lectures on the Microscope," vol. 36 (1888): 1149.

rically the principle of the plano-convex lens, noting that an object placed in a dense spherical medium between the eye and the center of the sphere would appear larger. Vitello in about 1278 and Roger Bacon in the *Opus majus* of 1267 (which was not, however, printed until 1733) discussed lenses, the latter claiming that it was possible to "shape transparent bodies and arrange them in such a way with respect to our sight and objects of vision" that "from an incredible distance we might read the smallest letters and number the smallest particles of dust or sand."[9] Eyeglasses for the old came into use around the beginning of the fourteenth century, having been independently invented in China where they were originally made of a dark transparent mineral.[10] In Europe, beryl (hence the German *Brillen*, spectacles), or rock crystal, rather than glass was used as a magnifier.[11] Nicholas Cusanus's *De beryllo* describes it as "a gleaming, bright, and transparent stone. To it is given a concave or convex form, and those who look through it discover things which were at first invisible."[12]

From these passages, it is plain that lenses might be considered wonderful on account of their ability to correct defective sight or their ability to extend the range of vision beyond the normal and natural. Unfortunately, the texts are ambiguous as to which is meant. Familiarity and rationalization worked against the ascription of magic, but even sixteenth-century references to lenses are difficult to interpret. Thus Robert Recorde alludes to Friar Bacon's glass "in which he might se thinges that wer doen in other places," which glass was thought to be given its powers by evil spirits, but whose capacities he, Recorde, knew to have been developed by means "good and naturall, and to be wrought by Geometrie . . . and to stand as well with reason, as to see your face in common glasse."[13] Remarks about devices enabling the user to see across vast distances have often been taken as indicating that sixteenth-century opticians were familiar with the telescope and understood its theory. Thomas Digges, in the preface to his father's 1571 work, the *Pantometria*, said that "by proportionall Glasses duely situate at convenient angles," Leonard Digges had been able to see as far as seven miles off, and the book describes the use of compound systems of concave and convex mirrors.[14] In book 17, entitled "Of strange glasses," of the second edition of his *Magia naturalia* of 1589, Giambattista della Porta says that by combining a concave lens (for seeing faraway objects

[9] Bacon, *Opus majus of Roger Bacon* 2:582.
[10] Rosen, "Invention of Eyeglasses."
[11] See Govi, "Compound Microscope," p. 583.
[12] Cusanus, *Ueber den Beryll*, p. 4.
[13] Recorde, *Pathway to Knowledge*, preface.
[14] Gunther, *Early Science in Oxford* 2:290.

clearly but small) with a convex lens (for seeing them magnified but dim), "you shall see things both afar off and things near hand, both neater and clearly." But as Albert van Helden has argued, most pre-seventeenth-century references to the powers of lenses to make the invisible visible are simply references to eyeglasses, rather than to visual aids like telescopes and microscopes, which provide for superhuman perception. Porta's description of a lens combination most likely refers to bifocal spectacles rather than to a telescope-microscope.[15]

The evidence points accordingly to the dating of both instruments—considered not as projections or fantasies, but as physically realized—to the first decade of the seventeenth century.[16] The Galilean telescope and the compound microscope each consist of a convex lens and a concave lens: the difference lies only in the separation of the two lenses, and suggestions that the compound microscope antedated the telescope rather than preceding it slightly are thought unconvincing by modern scholars.[17] The discovery that myopia was correctable with lenses of a strong negative curvature was reported in 1575 by Maurolycus. While presbyopia results almost inevitably from the aging process, myopia can be facilitated in its development by a lack of practice in focusing on distant objects. The disorder became widespread with the expansion of literacy and reading by juveniles. Although shortsightedness appears to have been recognized a century earlier as correctable by lenses,[18] the scarcity of concave lenses made experimentation with combinations of concave and convex lenses unlikely; the increase in the manufacture of and trade in spectacles in the last quarter of the sixteenth century made the discovery of the power of such combinations inevitable.[19] In 1608 the spectacle-maker Hans Lipperhey of Middelburg applied (unsuccessfully, for his priority was already contested) for a thirty-year exclusive right of manufacture for a set of lenses mounted in a tube. The resulting telescope, quickly transmitted westward into England and southward into France, Germany, and Italy, achieved instant celebrity through the series of sensational observations of the moon and planets made by Galileo in 1609–10 and reported in his *Sidereus nuncius.*[20]

The compound lens system qua microscope insinuated itself far more

---

[15] On the problem of the interpretation and assessment of optical passages in John Dee, Robert Recorde, Thomas Digges, Leonard Digges, and Giambattista della Porta, see Helden, *Invention of the Telescope*, pp. 12ff.

[16] Ibid., pp. 24–25.

[17] Ibid., p. 5 n. 1.

[18] According to Iliardi, 'Eyeglasses and Concave Lenses."

[19] Bell, *Telescope*, p. 2. The results of trial and error with different lens combinations are described by Helden, *Invention of the Telescope*, pp. 17–18.

[20] King, *History of the Telescope*, pp. 30ff.

gradually. There were several reasons for this. First, the magnification of small objects, as opposed to distant ones, was not a novelty and could be accomplished when desired with a single lens; second, the microscope was not presented as an agent of a revolution in physics and natural philosophy, as the telescope was by Galileo; third, being able to see far was obviously useful in affairs of the world—in navigation and warfare especially—but being able to see small was not. The technology was transferable. Galileo had used magnifying lenses for small objects even in the period of the *Sidereus nuncius* and later copied the design of the compound instrument,[21] and Cornelius Drebbel designed a microscope based on the principle of the Keplerian telescope, consisting of two convex lenses, and exhibited it in Paris in 1621 or 1622.[22] But microscopes were still something of a rarity in 1625.[23]

That less contention was attached to the question of the true inventor of the microscope than to the question of the true inventor of the telescope shows that it was the habilitation of the instrument, the creation of a context for it in which magnification had a purpose, that required originality and purposiveness. It is appropriate here to acknowledge the role played by Galileo's early promotion of (in contrast to Bacon's near indifference to) artificial optics as a means of achieving knowledge. The book that popularized the microscope, Hooke's *Micrographia*, did not appear until more than half a century after the *Sidereus nuncius*. It then both demonstrated the scope of microscopy—its application to the plant, animal, and mineral worlds, and to human artisanship—and supplied a moral-theological-practical justification for it.

For all its importance in securing a place for the microscope in the study of nature, *Micrographia* was not the first book of illustrations and descriptions of the microworld. To the early history of the genre belongs the collection of Joris (or George) Hoefnagel (or Huefnagel), who painted hundreds of insect studies in watercolor on vellum. These were copied by his son Jacob, who, in 1592, at the age of seventeen, produced an affecting series of compositions of plant parts, insects, and small animals in death or alive, embellished with mottoes on life, love, growth, and transience, in copperplate. The *Diversae insectarum volatilium icones* (1630) is presumably by the same artist, though it is less impressive, and the mottoes are dropped. Prince Cesi of the Lyncean Academy issued a large-format *Apiarum* in 1625, accompanied by an engraved plate of the bee and its parts as drawn by Francesco Stelluti,

[21] For references and documents, see Govi, "Compound Microscope."

[22] Ibid., p. 575; cf. Huygens, *Oeuvres complètes* 13:513.

[23] On the invention of the microscope, with chronologies and documents, see Waard, *De Uitvinding des Verrekijkers*; on its early history, see Disney, *Origin and Development*, pp. 89–115; on nomenclature, pp. 98–99.

who published a second version in an edition of poetry in 1630.[24] Giovanni Hodierna at Palermo followed with an illustrated treatise on the insect eye, *L'occhio della mosca*, in 1644, and the astronomer Francesco Fontana produced four pages of illustrations of insects in 1646.[25] Meanwhile Galileo had been carrying out microscopical observations with modified telescopes. In 1624 he was reported to be in possession of an instrument "of no greater height than a dining room table" that could multiply fifty thousand times, a figure that, given the convention of expressing power with reference to three dimensions, meant that it magnified by about thirty-six diameters, showing, as he said, a fly as big as a hen.[26] "I have contemplated many animals with infinite admiration," he says, "amongst which the flea is most horrible, the gnat and moth are most beautiful; and it was with great satisfaction that I have seen how flies and other little animals manage to walk sticking to the glass and even feet upward."[27]

Illustrations of microscopes themselves followed the publication of the first microscopical illustrations. Athanasius Kircher had microscopes in his possession by 1634 and published drawings of them in his *Ars magna lucis et umbrae* of 1646, under the chapter heading "De mira rerum naturalium constitutione per smicroscopium investiganda." He describes the "marvels concealed in the structure and organization of natural objects," such as the flea's foot.[28] Gaspar Schott, in his compendium *Magia universalis* of 1657–59, published a puzzling set of illustrations of six kinds of microscopes. Schott states that "many wonderful things hidden and concealed in past ages have been detected" by the microscope, but he does not emphasize their truth-revealing potential. Like Porta, he presents them in the context of natural magic and illusionism in the company of other devices for producing trick appearances: people like to fill a receptacle fitted with a lens with "very minute fragments of stones of various colours, with gold and silver filings, with little balls of coriander and suchlike trifling things. . . . If you now look into the vase you will think you see great pieces of stones, gold and silver twigs, pumpkins, rosy apples, and you will wonder."[29]

The savant Peiresc had reported that the microscope was still unfa-

[24] Stelluti, *Persio tradutto*.

[25] Fontana, *Novae coelestium terrestriumque rerum observationes*. On Hodierna's researches and their influence, see Fournier, "Fabric of Life," pp. 47ff.

[26] According to a letter of Peiresc, in Govi, "Compound Microscope," p. 576.

[27] Galileo, letter to Cesi, 23 September 1624, in Govi, "Compound Microscope," p. 577.

[28] Kircher, *Ars magna*, p. 834.

[29] Schott, *Magia universalis*, p. 534.

miliar in Rome in 1624.[30] But certainly by the time Hooke's book appeared in 1665, as we know from the *Journals* of Balthasar de Monconys, the arts of making and using lenses were widespread: Monconys found artisans and other aficionados wherever he went, in England, France, Italy, the Low Countries, and Germany. Richard Reeves's microscopes had reached the north of England by 1662, and Monconys found him at work in London in 1664.[31] Eustachio Divini and Giuseppe Campani were active in Rome, Johannes Wiesel in Augsburg, and the algebraist, statistician, and mayor Jan Hudde in Amsterdam during the same years; Vossius the Younger, the son of Descartes's old antagonist, was interested in light, color, and microscopes. Samuel Musschenbroek made them, including some used by Swammerdam, as part of his busy trade in machines and measuring instruments. Monconys found the Montmort circle in Paris busy with lenses in 1656. In 1670 the opticians' trade was flourishing in England, with manufacturers supplying, as Boyle said, "not only our own virtuosi but those of foreign countries with excellent microscopes and telescopes."[32] This community of manufacturers existed in an uneasy state of mutual dependence with the virtuosi themselves, on whose commissions they relied, but whose demands for openness and reciprocal exchange of technical data violated the tradition of craft secrecy.[33]

Kepler had given a summary of the theory of lenses in his *Ad vitellonem paralipomena* of 1604 and his *Dioptrice* of 1611. In the latter he gives a geometrical account of the workings of lenses and lens systems, enunciates the major optical theorems, and provides the solution to various optimization problems. By 1646, Monconys was able, as a result of the dissemination of this knowledge, to explain his designs for microscopes of four and five lenses, which, he said, were based on the principles enunciated in the *Dioptrique* of 1637 of Descartes and a chapter of Cavalieri's *Exercitationes geometricae* of 1647. Descartes's contribution to the subject, though he failed to give credit to Kepler and characteristically overstated his own originality and independence, was

[30] Extracts from his unpublished letters are reprinted in Govi, "Compound Microscope," p. 590.

[31] This was the origin of what A. D. C. Simpson refers to as the "scientific support community" of instrument makers; see Simpson, "Robert Hooke." The best contemporary survey of early microscopes was Louis Joblot's two-part work of 1718, *Descriptions et usages de plusiers nouveaux microscopes*, containing twenty-two plates of microscopes then in use, with their specifications and a record of his own experiments and observations.

[32] Boyle, *Works* (1744 ed.), 3:139.

[33] See the full social and technical study by Daumas, *Scientific Instruments*, and Simpson, "Robert Hooke."

substantial. Leibniz was being unusually sharp when he suggested that Descartes's optical theory and his designs for "magnifying glasses with which we could see animals on the moon and all the finer parts of creatures" were that philosopher's greatest and most substantial achievement, by contrast with his absurd and imaginary metaphysics.[34] But he understood the significance of Descartes's view of the relations between theory and practice. With the help of the mathematicians Mersenne, Mydorge, and Villebressieux and the technician Ferrier, and later with the help of Huygens,[35] Descartes summarized the theory of lenses and advanced it, dedicating himself to two problems: the optimal shape of a magnifying lens and the construction of a machine for grinding lenses. His aim was to transform an empirical craft into a rational discipline.

It is to the shame of the sciences, he says in the first *discours* of his *Dioptrique*, that the invention of magnifying lenses has been a matter of trial and error. Thirty years ago, on his account (which places the scene in 1607), a young man from a mathematical family who made mirrors and burning glasses, experimenting with different shapes, and who even worked with lenses of ice in the winter, happened to hold up a biconvex and a biconcave lens together and fix them at opposite ends of a tube. Though optics has been practiced and cultivated since, he complains, it has not been put under the control of theory and so has not progressed as it might have. His own treatise is intended to instruct artisans and their patrons.[36] Necessarily, then, it is written so as to be "intelligible to everyone."

In the tenth *discours*, Descartes describes the construction and use of single-lens microscopes "with which one . . . may see the divers mixtures [*mélanges*] and arrangements of small parts of which animals and plants and perhaps the other bodies which surround us are composed and from this draw many advantages for arriving at a knowledge of their natures."[37] He explains why a flea under the microscope looks as large as an elephant by saying that the image formed by the lens on our retina of a flea at a short focal distance is the same size as that of an elephant at thirty paces, and notes that "it is on this principle alone that the whole invention of little flea-glasses made of a single lens is founded, whose usage is sufficiently common everywhere, although no one up to now has understood the shape that these ought to have."[38] He then gives the optical principles of refraction, supplies figures, drawn by his assistant Schooten, of a simple and a compound microscope, shows

---

[34] Leibniz, *Philosophische Schriften* 4:298.
[35] Legrand, "L'invention des lunettes," p. 8.
[36] Ibid.
[37] Descartes, *Dioptrique*, in *Oeuvres* 6:3.
[38] Ibid., p. 155.

how to determine refractive indices, and, finally, explains how to construct a machine to grind lenses.[39] His most famous microscope design is shown as an instrument taller than its user, manned like a telescope.[40] It uses the sun and a reflecting mirror as a light source, and consists of a biconvex object lens and a plano-concave eye lens ground to hyperbolas; a parabolic mirror reflects light onto an opaque specimen, while a condensing lens is used underneath a transparent specimen. The difficulty with the conception, as Mayall points out, is that objects to be examined would have been instantly vaporized by the heat of the illumination source.[41] Like Schott's illustrations, this one is probably out of proportion.[42] Large microscopes were, however, produced by some makers, such as Divini. One of his models from 1671 has a tube extendible up to 56.5 centimeters, though the report that he had constructed a microscope with a tube as big as a man's leg and glasses as large as the palm of a man's hand was probably incorrect.[43]

Communication among opticians in Paris, the Netherlands, and Italy, and between those in England and on the Continent, is an important subject that has been little investigated. The northerners seemed, however, to specialize in handheld single-lens microscopes, which, unlike the common flea glass, offered high magnification, the southerners in tabletop compound systems. Monconys found Hudde in 1663 making little microscopes with a single lens by melting tiny glass beads in the flame of a lamp and polishing them with salt.[44] Hooke describes the process of making glass-bead microscopes in the *Micrographia*, and Isaac Vossius, too, was using a tiny hemispherical lens enclosed in a piece of wood with a tiny tube in front of it.[45] Huygens employed single lenses beginning in 1654.[46]

Leeuwenhoek's knowledge of and preference for the single-lens mi-

[39] The result, according to Christian Legrand, is "a veritable technico-scientific manual." "L'invention des lunettes," p. 15.

[40] Descartes, *Dioptrique*, in *Oeuvres* 6:207.

[41] Mayall, "Cantor Lectures on the Microscope," vol. 34 (1887): 994.

[42] Descartes was dissatisfied somewhat with the figures for the *Dioptrique*; see Legrand, "L'invention des lunettes," pp. 3ff.

[43] Adelmann, *Marcello Malpighi* 2:830.

[44] Monconys, *Journal des voyages* 2:161–62. The invention, however, may have been made by Torricelli, who produced "perline" by this means; see Govi, "Compound Microscope," p. 597. The method is a time-honored one: according to Harting, with some window glass, a platinum sheet, a spirit lamp, two tweezers, a small hammer, a few sewing needles, and some other household items, one can make a dozen such lenses in an hour with magnifications ranging from 80 to 2,000x. (Henry Baker, furnished with such a small glass bead, $\frac{1}{32}$" in diameter, which allegedly magnified 2,560x, said that he could see nothing.)

[45] Monconys, *Journal des voyages* 2:128.

[46] Huygens, *Oeuvres complètes* 8:48ff., 290ff.

croscope is not difficult to explain. The number of lenses possessed by a microscope is not a measure of its excellence as a research tool. Johann Zahn, in his *Oculus artificialis* of 1685–86, refers with contempt to "flea glasses," low-power tubes of a single lens with an eyepiece, which he calls *microscopia ludicra* or *curiosa*, as opposed to the compound *microscopia seria*. But the simple microscope, which might possess an extra condensing lens, was, for many applications, the instrument of choice. This type, which might be ground or made from a bead, was used by Leeuwenhoek from the 1670s until his death in 1723; it was probably the choice of Malpighi as well, though he also seems to have availed himself of several compound models.[47] According to Harting, glass-bead lenses will sometimes assume by chance an elliptical or hyperbolic surface, which reduces aberration: to hand-grind spherical lenses to such specifications is very difficult.

Gerald Turner states in his survey that "the resolution of the compound microscope remained virtually static from its invention in about 1600 to around 1830," when J. J. Lister published an account of lenses free of chromatic aberration.[48] One might wonder how "good" by modern standards microscopes of the second half of the seventeenth century were: the accompanying table should give some general orientation.[49]

For the sake of comparison, a modern oil-immersion microscope will magnify 600–1,000x; student instruments are usually outfitted with a

TABLE 1

Instrument	Magnification	Structures Visible
"Flea glass"	10x	Insect parts
Hooke's compound microscope	30–50x	Details of plant and animal sections, some pond organisms
Simple microscope	to 275x	Bacteria, protozoa, blood corpuscles
Other compound instruments	50–150x	Animal sections

[47] Adelmann, *Marcello Malpighi* 2:828ff.

[48] Turner, "Microscopical Communication," p. 5. See also Turner, "Microscope as a Technical Frontier."

[49] The following estimates are taken from Ford, *Single Lens*, pp. 127ff. See also his discussion of the capabilities of the surviving Leeuwenhoek instruments in *Leeuwenhoek Legacy*, pp. 51ff.

10x and a 40x objective lens and a 10x ocular lens. The optical limit of visibility is established by the nature of light and was shown by Helmholtz in the nineteenth century to be .2 micrometers (.2 μm). A 300x lens can, under good conditions, resolve to 1 micrometer. A human hair is about 100 μm and plant cells are about 50 μm in diameter; the latter could thus be seen even by Hooke with his low-power compound instrument. Blood cells are, however, about 7–8 μm in diameter, and bacteria are 2–5 μm, within reach of the single-lens but not the compound microscope of the period.

Power of magnification alone was not the hindrance to the development of microscopical science that might be imagined; the resolutive capacity of a microscope that enables it to turn a large blur into two distinct points is a better measure of its capacity. A Leeuwenhoek microscope could in fact resolve, according to Turner, to 1.25 μm. But the design of an optical instrument is essentially a matter of compromises. Geometrical optics itself is a straightforward theory, much of which can be deduced from two propositions: first, that light travels in straight lines in a homogeneous medium; second, that in passing from one medium into another of a different density, rays are refracted. Accordingly, with a knowledge of the refractive index of the material used to fashion the lens, and a specification of its shape, the properties of the lens should be fully determinable and optimal results achievable.[50] In practice, however, there are obstacles to perfection. Increase the magnifying power of the lens, or of the microscope by compounding lenses, and the amount of light entering the eye is decreased correspondingly: with magnification of 20x, light is decreased by a factor of twenty, the reason for the "dark and gloomy views" of which early users of the compound microscope complained.[51]

A lens produces a curved image; either the edges are out of focus or the center is. This can be compensated for by using a curved surface to mount the specimen, but then not all of its parts will be equally enlarged. By making the lens opening smaller, one achieves some improvement, but again at a cost of light. Chromatic aberration is a consequence of the fact that different wavelengths of light are differently refrangible; the red rays arrive in a different location from the blue rays and produce colored halos or fringes around the edge of the image. Newton had argued that chromatic aberration could not be reduced on

[50] Ian Hacking has emphasized recently that the theory of geometric optics is not the correct theory to explain the magnification of the light microscope, which depends on the effects of diffraction. Hacking, "Do We See through a Microscope?" p. 306. Nevertheless, it "saves the phenomena" of magnification.

[51] For a generally negative assessment of early microscopes, see Bradbury, "Quality of the Image," and Turner, "Microscope as a Technical Frontier."

account of the dispersiveness of all transparent bodies. The aberration can be corrected by using two kinds of glass of different refractive indices; but the theory of the achromatic refracting telescope was published only in 1758,[52] and its manufacture was not perfected until much later. Finally, not one ray of light but a multitude depart from every point on the observed object and are focused differently; this is responsible for the blur known as "spherical aberration." Elliptical and hyperbolic lenses, as Descartes saw, will also eliminate spherical aberration, but these are exceedingly difficult for a technician to grind and were beyond the reach of seventeenth-century opticians.[53]

Aberration was a far more serious problem for compound than for single microscopes. Yet the compound microscope had certain advantages, in addition to its larger field. The lenses were not so minute and difficult to handle: among the single lenses made for the Royal Society by John Mellin in about 1680, for example, was one only one twenty-fifth of an inch across.[54] The compound instrument could be constructed with a stand and made to sit on a table, leaving the hands of the viewer free to manipulate the specimen, and it could be fitted with interchangeable objective lenses to give varying powers of magnification in conjunction with the ocular lens. It could be attractively decorated, as Hooke's tooled and gilded leather-covered instrument was. Though Leeuwenhoek used silver and sometimes gold for the body of his microscopes, their beauty was and is recherché compared with the gleaming proportions of the larger instruments. The compound microscope did not have to be positioned so close to the eye, which many observers found uncomfortable. As Hooke complained:

> the smaller the sphere is in which they are made, the nearer do they bring the object to the eye; and consequently the more is the object magnified . . . but to make any Sphere less than 1/10 of an inch in Diameter is exceeding difficult, by reason that the glass becomes too small to be tractable; and 'tis very difficult to find a cement that will hold it fast whilst it be completed; and when 'tis polisht, 'tis exceeding difficult to handle and put into its cell: besides, I have found the use of them offensive to my eye, and to have much strained and weakened the sight, which was the reason I omitted to make use of them, though in truth they do make the object appear much more clear and distinct and magnifie as much as the double Microscopes.[55]

But Hooke testified to the superiority of the single lens "to those whose eyes can endure it," observing that the colored fringes that disturb the

[52] An account is given in Dolland, "Some account."

[53] Descartes, *Dioptrique*, in *Oeuvres* 6:199ff.

[54] Clay and Court, *History of the Microscope*, p. 44.

[55] Hooke, *Cometa et microscopium*, p. 96.

image in the compound microscope are avoided with the single lens.[56] There were other inconveniences experienced with single-lens microscopes: the short focal length of high magnifiers meant that the object had to be placed extremely close to, or even touching, the lens, dirtying it or sometimes scratching it. Holding the apparatus up to the light was tiring, and mounting the specimen by, for example, getting it to stick on a pin was more trouble than having it lie flat. For all these reasons, an optically inferior device might be preferred. "I have had *Mellens's* glasses, and seen *Leeuwenhoek's* and *Campani's,*" says the author of the *Lexicon technicum* of 1710, "but I would sooner have the double microscope than any of them . . . and the price is much easier."[57] But even in the middle of the eighteenth century, John Hill could claim that the compound microscope was a "plaything . . . for those who would be diverted by the power of magnifying" and that all serious discoveries had been made with the single lens.[58]

Both the single and the double microscope are focused by bringing the object to the correct distance. The early Hooke-type microscope required the user to push or twist an inner tube into or out of an outer one, which was not easy to do without jarring the specimen. Screw-driven focusing, added by Johann Hevelius in 1673, was thus a major improvement. Other modifications, such as the substage mirror introduced by Edmund Culpeper in 1725, addressed the problem of illumination. English opticians understood the contribution of a condenser, often shown in illustrations, either separate from the microscope or in its modern substage position. This took the form of a tube or globe filled with water or brine, with or without additional lenses; it focused light on the object. The ball-and-socket joint for swiveling the apparatus was also pioneered by Hooke, and his experimentation with various refractive agents was broad: he tried out "waters, gems, resins, salts, Arsenick, Oyls, and . . . divers other mixtures of watery and oyley liquors." Gemstones—diamonds, sapphires, rubies, beryls, and topazes—were used in place of glass lenses in the hope of achieving distortion-free magnification. Other early modifications added to the versatility of the instrument: the solar microscope of Johann Nathaniel Lieberkuehn allowed an image to be projected on a wall; John Cuff's water microscope permitted free-swimming creatures to be examined in their element; and "universal microscopes" put a range of lenses on a rotating wheel. Cherubin d'Orleans insisted on the superiority of the

[56] Ibid.

[57] Harris, "Microscope," in *Lexicon technicum* 1:513.

[58] Hill, *Essays*, p. 124. For an elaboration of this instrument's virtues, see Ford, *Single Lens*, pp. 121ff.

binocular microscope, the design for which he introduced in *La vision parfaite* of 1677.

An irritating feature of microscopes is that the image appears upside down, so that moving the slide around merely to get the object in view is a frustrating exercise for the novice. This presented less of a problem in the compound microscopes sold to amateurs, which used prepared slides along which objects were ranged in a strip, but these were highly restricted as to subject. Other obstacles to successful observation are the fault of the specimen. Hooke saw that the dry-mounting of anatomical segments—nerves, muscles, tendons, ligaments—revealed little or nothing; the specimen had to swim in oil or water to show its structure. It had a tendency nevertheless to dry up in the presence of heat and light, and it had to be cut in thin sections in order not to appear under high magnification simply as a dark blob. The microtome, a screw-driven apparatus that lets thin wax-coated specimens be sliced off precisely, was not used until 1770; before then one was thrown back on the use of knives and scissors.[59] Boyle found that it helped to freeze the specimen first, though tissue deteriorates rapidly on thawing. Another difficulty is the transparency of many plant and animal parts and of microorganisms, which are invisible without stains. Some natural dyes were tried—Leeuwenhoek used a solution of saffron in brandy to study transparent muscle fibers—but before the introduction of synthetics such as the basic eosin red and methylene blue in the nineteenth century, the problem was a serious one.

The range of designs of the earliest microscopes, the commercial varieties of which all issued from a few well-known workshops, contrasts with the progressive standardization the instrument underwent as trial and error proved some designs more functional than others and as aesthetic considerations became less and less relevant to opticians in the design of scientific instruments. Price is correct to emphasize that instruments experience a symbolic-decorative phase before utilitarian values gain the upper hand entirely.[60] One might accordingly suspect that, as Bracegirdle suggests, the scientific community in the seventeenth and eighteenth centuries was more interested in designing and possessing optical instruments than in using them to explore the world.

Here, though, is another example of perceived delay: the motives of the users and the optical quality of early instruments are questioned, and it is argued somewhat inconsistently by scholars both that early instruments were too defective to produce worthwhile results and that

---

[59] On the problems of sectioning, see Ford, *Single Lens*, pp. 41ff. Ford discovered and reexamined several of Leeuwenhoek's original specimens, which he describes elsewhere.

[60] Price, "Of Sealing Wax and String."

the conceptual limitations of contemporary observers prevented them from taking full advantage of the optical resources available to them. But in fact an explosively wide range of investigations was made in the fifty years between 1640 and 1690. Leeuwenhoek looked at semen, blood, milk, bone, hair, spittle, the brain, sweat, fat, tears, sap, salts and crystals, protozoa and parasites, sponges, mollusks, fish, spermatozoa and embryos, pores and sweat, and muscle fibers. Grew specialized in plant parts; Malpighi, the most methodical, studied the kidneys, the lungs, the gall, the brain, fat, and bone marrow, as well as chicken embryos and the fine anatomy of insects, particularly the silkworm. Swammerdam, in his short career, was drawn mainly to insect anatomy and metamorphosis.[61]

The first subjects of observation and controlled description were, as noted, insects. Given what Descartes says about the popularity of flea glasses, and Monconys's habit of carrying his about with him, it might seem inevitable that the pastime of observation should inspire a written record. That the subjects of observation were lowly and ubiquitous rather than precious and rare was offset by the novelty of their magnified appearance. The first truly systematic illustrated study is that of Pierre Borel, whose *De vero telescopii inventores* (1655), which deals unsurprisingly with both microscopes and telescopes, was usually bound together with his *Centuria observationum microscopicarium* (1656). Borel saw and sketched, crudely, seeds, insects, and bodily fluids, including blood and semen. This work was followed by the better-known but unillustrated study by Henry Power in 1664 and by Hooke's classic, with the beautiful illustrations probably done by Christopher Wren.

Power's *Experimental Philosophy* continues the tradition of nature observation practiced by Thomas Moffett, to whom he often refers, and it shows his debt to his mentor, Sir Thomas Browne, who was a lover of pattern, form, and miniaturization. A powerful effort is made to win the reader over through the charm and human interest of the presentation: the three naturalists ruthlessly exploit the attraction noted by Bachelard, which depends for its effect on the possibility of comparisons between objects of the macroworld and those of the microworld, on a premise of reduplication. None of these works is provided with illustrations, so the author's prose must bear the entire burden of communication, leading to a certain archness of style. Typical is Power's description of the flea:

> It seems as big as a little prawn or shrimp, with a small head, but in it two
> fair eyes, globular and prominent of the circumference of a spangle. . . . He

[61] Their individual contributions are effectively detailed for the first time chapter by chapter by Fournier, "Fabric of Life."

has also a very long neck, jemmar'd like the tail of a Lobstar, which he could nimbly move any way; his head, body, and limbs also, be all of blackish armour-work, shining and polished with jemmar's, most excellently contrived for the nimble motion of all the parts: Nature having armed him thus *Cap a pe* like a Curiazier in warr. . . . His neck, body, and limbs are also all beset with hairs and bristles, like so many Turnpikes, as if his armour was pal-ysado'd about with them. At his snout is fixed a Proboscis, or hollow trunk or probe, by which he both punches the skin and sucks the blood through it.[62]

Power looks at the bee and at "eels" in vinegar, discovering that they are killed by heat and by vitriol, but not by cold. After a freeze and a thaw, "all my little Animals made their reappearance, and danced and frisked about as lively as ever."[63] He notes that the smallest animals, those that seem most to resemble plants, have the most restless motions, and he projects this feature to the inanimate world. "[T]he innumerable number and complicated motion of these minute animals in Vinegar, may very nearly illustrate the Doctrine of the incomparable Descartes, touching Fluidity, (viz.) That the particles of all fluid bodies are in a continual and restless motion . . . notwithstanding that the unassisted eye can discover no such matter."[64] He examines lines drawn on paper, cloth, metal, and powder, as well as pollen, leaves, and sparks. There is an opportunity for a nice conceit in the notion of an eye observing, with the help of a lens based on conic sections, the insect eye with its own conic architecture.

It has been observed that, in the pre-Newtonian phase of the Royal Society, by contrast with its eighteenth-century phase, living subjects occupied more attention than celestial mechanics; Boyle even went so far as to say that the contrivance of the eye of a fly or a man's muscles was superior to the sun and the system of the heavenly orbs.[65] It was not simply a question here of the prestige of the mathematical sciences after the *Principia* and Newton's ascendancy, but one of positive avoidance. There were theological reasons in the earlier period for preferring to concentrate not on lumps of matter in infinite spaces but on exquisite contrivances.

This interest was amply repaid: *Micrographia* is a captivating book. Its author, Hooke, who was apparently the brains behind the experi-mental, mathematical, and mechanical competence of Robert Boyle, the great rhetorician of the mechanical philosophy, was first employed di-

---

[62] Power, *Experimental Philosophy*, pp. 1–2.
[63] Ibid., p. 35.
[64] Ibid., p. 36.
[65] See Espinasse, *Robert Hooke*, p. 28.

rectly by him when Boyle went to Oxford in 1654.[66] In 1662 he became the paid curator to the infant Royal Society, and acquired other titles in due course: city surveyor, Cutlerian Lecturer in Mechanics, and Gresham Professor of Geometry.[67] He took over the making of microscopical observations from Christopher Wren in 1661, the society having ordered him to present and publish observations on moss, vinegar, bark, blue mold, and spiders' eyes. He brought his microscope to Royal Society meetings, mainly during 1663 and 1664, and on those occasions invited the members to compare his "schemes" or drawings of those subjects with their appearance through the microscope. These drawings were composites, not records of individual observations. "I never began to make any draught," he says, "before by many examinations in several lights, and in several positions to these lights, I had discovr'd the true form."[68] For Hooke, these illustrations were not meant to be direct reproductions of a momentary optical experience, but rather an improvement on momentary witnessing that would give the general form stripped of the idiosyncrasies of the individual specimen or observation.

Hooke's objects of study were insects, especially fleas, lice, flies, moths, and bees; plant material; mold and fungus; inorganic objects, such as snowflakes and stones; man-made objects, such as needles, razors, cloth, and paper; animal parts, such as hair, fish scales, and the sting of the bee; and transitory phenomena, such as sparks struck from flint and the colors of thin films of mica. Observation 18 is the famous examination of cork with its "little boxes or cells" without passageways between them, which Hooke supposed to be an excrescence on the tree's bark. *Micrographia* was meant, it has been argued, as an offering to the Royal Society's sponsor, King James, as evidence of the accomplishments of the society.[69] It upheld, as John Harwood points out, the society's claims to be directly concerned with "things," to be dedicated to an ideal of objectivity, and if what *Micrographia* delivered was not precisely useful knowledge, the society had at least shown itself capable of producing something other than talk and socializing. The book was vastly more entertaining than the artificial-language-manual-cum-encyclopedia that the society had recently sponsored in John Wilkins's dinosauric *Essay towards a Real Character.* But Hooke's most prescient statements are deeply buried in the observations, where they are often tangential to the ostensible subject matter; Hooke used his assignment to communicate some of his freest speculations on physics in general.[70]

[66] On Hooke's career, see ibid., esp. chap. 3, and Hunter and Schaffer, *Robert Hooke.*
[67] Espinasse, *Robert Hooke*, pp. 43ff.
[68] Hooke, *Micrographia*, preface.
[69] Harwood, "Rhetoric and Graphics in *Micrographia.*"
[70] Espinasse, *Robert Hooke*, pp. 50ff.

Into this category fall his remarks on colors and diffraction in observation 58, his statements about combustion, heat, and respiration in the section on charcoal (observation 16), his remarks about the cause of petrification (observation 17), and the aforementioned discovery of "cells" in cork. Yet the associated illustrations were not the most striking or memorable, and the prize for visual interest must go to the animal and vegetable sections. Newton, whose eye was for the serious, seems to have found most interesting the observations on plate colors and diffraction.

The reception of the book was highly positive: it was much sought after, and print runs sold out repeatedly. A summary in German appeared in 1667 with copies of the most popular engravings, and the *Journal des savants* reviewed it in 1668, lamenting that few people could read English. The *Journal des savants* praised the illustrations and included a copy of the magnificent louse foldout. Huygens wrote to Robert Moray of the Royal Society that he was delighted to have received the book at last, that he admired the care with which it had been drawn and engraved, the revelations of mechanics and geometry in the work of nature, and the theoretical studies of colors.[71] Leibniz had also heard of it and tried persistently for years to obtain a copy, receiving one finally in 1678.

It was in 1673, some years after the publication of *Micrographia*, that the Royal Society received its first communication from Leeuwenhoek, the Delft shopkeeper who lived from 1632 to 1723 and who made microscopical investigations continuously from the early 1670s to his death at the age of eighty-five. The Leyden anatomist Regnier de Graaf, who became famous for his own anatomical studies of the female reproductive system, had forwarded to the society's secretary, Henry Oldenburg, a letter in which Leeuwenhoek criticized some of the drawings in Hooke's book, taking particular issue with the representation of the bee's eye; de Graaf endorsed him as a microscopist superior to the well-known instrument maker Eustacio Divini. Leeuwenhoek followed up with other letters over the next few years but first made a solid impact with a communication of October 1676, read to the society the following February and partially published in March. It begins as follows:

> In the year 1675, about half-way through September ... I discovered new living creatures in rain, which had stood but a few days in a new tub, that was painted blue within. This observation provoked me to investigate this water more narrowly; and especially because these little animals were, to my eye, more than ten thousand times smaller than the animalcule which Swam-

[71] Huygens, letter to Moray, 27 March 1665, in *Oeuvres complètes* 5:281.

merdam has portrayed and called by the name of Water-flea or Water-louse, which you can see alive and moving in water with the bare eye.[72]

This was not the first account of very small animals, which Leeuwenhoek calls here "living Atoms." Divini had been reported in the *Philosophical Transactions* of 1668 as having found an "animal with many feet, its back white and scaly" and of very small size: as a grain of sand is to a nut, so was this insect to a grain of sand, which prompted the observers, he said, to describe it as the "*Atome of Animals*."[73] Leeuwenhoek, however, did not see a single mitelike animal, but whole colonies of creatures. These had the form of globules, some with thick tails, some with little legs, or fins, or "paws," some serpentlike, some so small he could not determine their shape, others as large as cheese mites, some of which exploded when they were taken out of the water. These were far smaller than Divini's creature: "as the size of a full grown animalcule in the water is to that of a mite, so is the size of a honeybee to that of a horse; for the circumference of one of these same little animalcules is not so great as the thickness of a hair on a mite."[74]

These animalcula appeared after a short time in water left exposed to the air, in clear well-water, and even in seawater, but Leeuwenhoek could find no animalcula in three-year-old snow water that had stood in a stoppered bottle. When unstoppered and mixed with pepper on 6 April 1676, this water failed to produce animalcula in the first week, and the water evaporated.[75] Fresh snow water added to the mixture brought nothing on 4 and 5 May, but on 6 May they suddenly appeared. "No more pleasant sight has ever come before my eyes," he says.

The Royal Society subsequently attempted to reproduce these observations, but the first trial, using "common pump water" in "exceeding small and exceeding thin Pipes of Glass of various sizes, some ten times as big as the hair of a man's head," failed on 1 November 1677. It was ordered that a proper infusion of pepper be prepared and that better microscopes be used, "that the truth of Mr. Leeuwenhoek's Assertions might, if possible, be experimentally examined, of which he had produced so many Testimonies, from such as affirmed themselves Eye Witnesses."[76] The eight witnesses assembled by Leeuwenhoek had included

[72] Leeuwenhoek, "Observations concerning Little Animals," in Dobell, *Antony van Leeuwenhoek*, pp. 117–18.

[73] Hooke, "Another Extract," reported also in *Journal des savants*, 12 November 1668, p. 108.

[74] Leeuwenhoek, "Observations concerning Little Animals," in Dobell, *Antony van Leeuwenhoek*, p. 123.

[75] Ibid., p. 133.

[76] Gunther, *Early Science in Oxford* 7:446.

"2 ministers, [a] Public Notary, and other persons of good Credit," and they all seem to have agreed that they saw little animals, though their estimates of the number in a drop of water varied from ten thousand to forty-five thousand. The next week brought nothing better. Though the microscopes provided by Hooke were judged "more convenient and expeditious," no animals could be seen. Vice President Henshaw thought that it might not be the right season for the breeding of these small insects, but Dr. Whistler took the more skeptical view that "these small imagined creatures might be nothing else than the small particles of the Pepper swimming in the water and no Insects."[77] (At the same meeting, William Croone proposed that a chicken might be seen formed in the cicatricula of an egg with the help of a microscope and was instructed to conduct observations accordingly.)

Finally, on 15 November, the experiment succeeded. A glass magnifying one hundred thousand in bulk (i.e., 100x) was used on pepper water that had been left undisturbed for nine or ten days, and some tiny things were seen that "by all who saw them were verily believed to be animals; and that there could be no fallacy in the appearance."[78] A list of witnesses was appended.

Leeuwenhoek's eagerness to make public his discoveries was coupled with a reluctance to inform his correspondents of his procedures. "My method for seeing the very smallest animalcules and the little eels [in pepper water]," he says, "I do not impart to others; nor yet that for seeing very many animalcules all at once; but I keep that for myself alone."[79] He provides, however, a detailed explanation of how he calculated the size and number of the animals he saw. Though he sometimes resorts to imprecise relative comparisons such as "eight times smaller than the eye of a Louse," he often uses a grain of sand as his standard. As he explains it, if there are one thousand little animals in a quantity of water the size of a grain of sand, and "if the *axis* of a grain of Sand be 1, the axis of a drop of water is at least 10, and consequently a drop is 1000 times bigger than that sand, and therefore [there are] 1000000 living Creatures in one drop of water." One counts a thousand animals the way one would a large flock of sheep running in a disorderly herd, by quickly arriving at an estimate of the length and breadth of the whole flock.[80]

Estimating size was still a problem, for the power of a lens could only be deduced from its magnification and could not be known indepen-

[77] Birch, *History of the Royal Society* 3:349.

[78] Gunther, *Early Science in Oxford* 7:450.

[79] Leeuwenhoek, "Observations concerning Little Animals," in Dobell, *Antony van Leeuwenhoek*, p. 144.

[80] Leeuwenhoek, "Monsieur Leeuwenhoecks Letter," p. 845.

dently. Hooke had hit upon the idea, which he describes in *Micrographia*, of looking down the microscope tube with one eye and at a ruler with the other. By this means he was able, he says, "to cast, as it were, the magnified appearance of the object upon the Ruler, and thereby to measure the Diameter it appears through the glass, which, being compared with the Diameter it appears to the naked eye, will easily afford the quantity of its magnifying."[81] Such an approach would not work, however, when the object was invisible to the naked eye to start; here something like Leeuwenhoek's method had to be used. A diminutive measuring apparatus was invented in 1718 by Jacobo Jurin. Jurin wound a wire around a pin; the diameter of a wind could be calculated as the length of the wire divided by the number of winds, and the apparatus was then placed in the field of view.[82] A grid fitted into the eyepiece was not used until 1732.[83]

In his *Cometa et microscopium* of 1678, Hooke gives detailed and easy-to-follow instructions for making single-lens glass-bead microscopes and for assembling compound instruments, along with the name of a reliable manufacturer.[84] Leeuwenhoek had told the society a year earlier in response to complaints about the difficulty of seeing his "little animals" and requests for better information, "The make of Microscopes employed by me, I cannot yet communicate."[85] This attitude was seen as uncollegial and insulting to the society's members. Their guiding principle was the free flow of information; they had tried to break down the mystique of the trade secret and to persuade craftspeople to communicate their methods in exchange for the public recognition and publication they had, as a society, to offer, and Leeuwenhoek did not seem to appreciate this. The fault was not entirely his; their own incentives were less than overwhelming. They did not—indeed, could not—make the generous publication offers they later did to Marcello Malpighi. Leeuwenhoek's rambling accounts, though they were collected in 1695 as the first edition of *Arcana natura*, did not exactly constitute a treatise or make a book, so that he could not perhaps have expected to gain much by adopting a more cooperative attitude. He seems to have seen that, because he lacked education, breeding, and a footing in the learned world, a revelation of his methods would have made him superfluous. So he gave his observations freely, in exhausting, even gratu-

---

[81] Hooke, *Micrographia*, p. 22.

[82] Jacobo Jurin, *Dissertations*, in Baker, *Microscope Made Easy*, pp. 45–48.

[83] This was the invention of the optician Martin, according to Miall, *Early Naturalists*, p. 203.

[84] On Hooke's relations with London instrument makers, see Simpson, "Robert Hooke and Practical Optics."

[85] Leeuwenhoek, "Monsieur Leeuwenhoecks Letter," p. 845.

itous, detail, but his methods, his best lenses, and his techniques were his capital, and he guarded them well. Failing to understand the importance the society attached to the reproduction of experiments, he was irate at the suggestion that anyone could not accept his accounts as accurate, but it was better in his view to communicate testimonials from reliable witnesses backing him up than to let his methods become common property. Even Hooke's offer of membership in the society in 1680[86] did not cause Leeuwenhoek to release better information. Leeuwenhoek's jealousy was described by an acquaintance of Constantijn Huygens, who reported on an unpleasant visit in 1685: Leeuwenhoek "did not want to show him any microscopes except those he shows to everybody," saying haughtily that if he ever did so he would soon become a slave to the whole world, and he refused to leave his visitor alone for a moment with even the public collection, saying over and over that he did not trust people, particularly Germans. "O what a beast!"[87]

The compiler Johann Zahn pointed out that Leeuwenhoek alone of the great manufacturers, among whom he included Samuel van Musschenbroek, Bonanus (Filippo Buonanni), and Hooke, had given no details of his microscopes, and Henry Baker, who examined the collection he left to the Royal Society, even came to the conclusion that Leeuwenhoek had destroyed his best microscopes before his death.[88] He understood that Leeuwenhoek had not in fact made a separate microscope for each specimen, and that the permanent fixtures were only a feature of the display microscopes. In May 1688, Fatio de Duillier, the sometime protégé of Newton, had already refuted in theory something of the Leeuwenhoek legend; Leeuwenhoek could not, he pointed out, have had a particular secret for getting high magnification, as there was a limit to the focal length one could give to a single lens.

Leeuwenhoek's publicizing of his discoveries brought him considerable celebrity. He was visited by such personages as Charles II, George I, Queen Anne, and Peter I of Russia, but—he complained—with no profit to himself. Leibniz and Huygens were surprised that he had not attracted patronage and financial reward, and Leibniz suggested that he open up a school. This proposal was not well received; Leeuwenhoek did not want to teach others his techniques and replied simply that the school idea had been tried by others and had never succeeded, as so few people had any real facility for this kind of research. It is inappropriate nevertheless to describe Leeuwenhoek as an amateur of science or to see him as only a clever craftsperson and good recording machine. His

---

[86] Birch, *History of the Royal Society* 4:6.
[87] Huygens, *Oeuvres complètes* 9:2408.
[88] See below, p. 225.

powers of concentration and his application to his subject stood, like Malpighi's and Swammerdam's, in contrast to the easy distractibility of the Royal Society luminaries. He theorized constantly on matters of importance: on the growth of living things, and on reproduction and voluntary movement. He discovered that the flesh of a pear consists of a multitude of tiny channels and speculated that the growth of the fruit depends on the flow of sap and the deposition of nutritive materials within these channels. He studied the muscle fibers of animal flesh in contracted and relaxed states and showed how movement was possible. He decided that bees and all other insects were machines that reacted automatically to patterns of light falling on their eyes, and, after confirming the discovery of a Dutch student, Johan Ham van Arnhem, of spermatozoa in human semen, he defended, with some modifications as time went on, an animalculist theory of generation to the end. The self-abasing tone he adopted with the Royal Society was mixed with a sort of plebian defiance. "I have oft-times been besought, by divers gentlemen, to set down what I have beheld through my newly invented *Microscope*: but I have generally declined; first, because I have no style, or pen, wherewith to express my thoughts properly; secondly because I have not been brought up to languages or arts, but only to business; and in the third place because I do not gladly suffer contradiction or censure from others."[89] Though he was coaxed into prolixity, his personality did not improve. Thomas Molyneux, brother of the optician and mathematician William Molyneux, who visited him in 1685, formed, like Huygens's Landgrave, a bad impression, finding him stubborn and pigheaded. "Leeuwenhoek knows nothing of the ideas of others, but he has such confidence in his own that he throws himself into extravagances or into bizarre explanations entirely irreconcilable with the truth."[90]

Leeuwenhoek's microscopes were powerful enough to enable him to see bacteria and some cells. He provided a sketch of cocci and bacilli from the human mouth with letter 39 to the Royal Society in September 1683. These he found in an old man who had lived a sober life and never drank brandy or took tobacco, and in another old man too, one who drank wine and brandy and whose teeth were "uncommon foul."[91] He did not, however, have a monopoly on high magnification. In the third part of his *Dioptrique*, "De telescopiis et microscopiis," Christiaan Huygens provided numerous rough sketches and brief descriptions

[89] Leeuwenhoek, letter no. 2, 15 August 1673, in Dobell, *Antony van Leeuwenhoek*, p. 42.

[90] Blanchard, "Les premiers observateurs," p. 413.

[91] Leeuwenhoek, letter no. 39, 17 September 1683, to F. Aston, in Dobell, *Antony van Leeuwenhoek*, pp. 238ff.

of infusoria, flagellates, paramecia, vorticella, monads, and spermatozoa, and what look like coccal bacteria, based on observations made
by him between 1678 and 1680.[92] He observed conjugation repeatedly,
but apparently not the fission of mother organisms. These pages are not
a contribution to an illustrated micrography—they were evidently produced in a hurry and were not intended as additions to this genre—but
they are to the literature of minute natural history. They are proof that
Leeuwenhoek was not uniquely in possession of excellent instruments;
research into remote forms of life was within the reach of many observers, including John Harris and Edmund King, who published accounts of protozoa in the *Philosophical Transactions* of the 1690s,[93]
and Filippo Buonanni, author of *Observationes circa viventia . . . cum
micrographia curiosa*, published in Rome in 1691. Leeuwenhoek excelled only in his single-mindedness, in the novelty of his results, and in
his willingness to theorize extensively about them.

Except in their encouragement of the student of plant anatomy,
Nehemiah Grew, who only began to use a lens, he tells us, after stimulated to do so by Malpighi, the Royal Society seemed to prefer to leave
sustained observation to foreigners. They farmed it out to Malpighi,
who had come to Henry Oldenburg's attention in 1667 and was made a
fellow the following year. Oldenburg commissioned Malpighi's silkworm treatise *De bombyce* in 1669 and arranged for the publication of
his other main anatomical studies.

Malpighi had left a logic post in Bologna to become professor of
theoretical medicine at the University of Pisa in 1656, where he learned
some mathematics, met Giovanni Alfonso Borelli, and participated in
the Academy of Cimento. He returned to Bologna in 1659 and remained there until 1691, except for an interval of four years from 1662
to 1666, before going to Rome, where he died in 1694.[94] His first work,
De pulmonibus, published in 1661, showed that the lung, one of the
"parenchymata" (literally, the "poured in beside"), was not merely
"fleshy," but had a distinctive microstructure of inflatable membranous
sacs of alveoli surrounded by blood vessels. In the living frog's lung he
saw the blood streaming through smaller and smaller branches of blood
vessels, which then reunited in a larger vein, thus demonstrating the
capillaries suspected by defenders of the circulation. Examining the
dried lung of a frog with a microscope, he was able to confirm that the
blood remained constantly within the vessels and did not seep out into

[92] Huygens, "De telescopiis et microscopiis," in *Oeuvres complètes* 13.2:690–713.

[93] Harris, "Some Micrographical Observations"; King, "Several Observations and Experiments."

[94] Different dates are supplied by Adelmann, *Marcello Malpighi*; the *Dictionary of Scientific Biography*, pp. 62–63; and Miall, *Early Naturalists*, pp. 146–47.

the surrounding flesh. In 1663 he began to study plants; he was allegedly made curious by seeing through a microscope that the threads projecting from a broken chestnut bough were really hollow air ducts, which he interpreted as analogous to the tracheas of invertebrates and the lung sacs of vertebrates. A preliminary version of a treatise on the anatomy of plants was sent to the Royal Society in 1671; the two parts of what became the *Anatome plantarum* followed in 1675 and 1679.

Malpighi's most sustained interest seems to have been in the structure of the internal organs of mammals, especially the kidney, brain, and uterus, as described in *De viscerum structura* of 1666. But at Messina he studied sea creatures, finding the pleated optic nerve of the swordfish particularly intriguing, and he was also fascinated by the microsurface of the tongue, which enabled him to propose a mechanistic theory of taste reception, according to which the papillae under the skin were stimulated by particles dissolved in the saliva. He examined the brain and spinal marrow, interpreting the brain as a gland, but found himself unable to explain the function of the brain from its structure, wondering whether it was really responsible for wisdom and intelligence in addition to sensation.[95] He saw the nervous continuity between the cortex and the periphery, in virtue of which the body was one sensitive system, though he expressed doubts about the notion of Descartes and Thomas Willis that liquid spirit could flow through the hollow nerves from the periphery into the brain. However, he was ready to follow Descartes in imagining glands as essentially sieves that separated particles of particular forms. Studying the growth of horns, of galls, and of warts as examples of normal and abnormal excrescences, he interpreted that growth as resulting from mechanical stimulation, and when he began to study plants, it occurred to him that there must be an analogy between the formation of woody stems and the growth of bones and teeth, which seemed to be produced in the embryo from filaments.

Oldenburg first wrote to him in October 1667, praising what he had heard of his work, and inviting him to convey his results to the society for publication. Malpighi was honored by the request and composed, in less than a year and a half, his classic treatise on the anatomy of the economically significant silkworm, with pictures and descriptions of its head, mandibles, and sexual organs, its spiracles and tracheas (which he saw served a respiratory function), its heart, fat, intestine, and rectum. His study of its development was followed by his equally celebrated study of the chicken egg, published as a sixteen-page essay, *De formatione pulli in ovo*, in 1673 by the Royal Society.

---

[95] Malpighi, *De cerebri cortici*, in Adelmann, *Marcello Malpighi* 1:302.

Seeking ever simpler structures, Malpighi turned to the anatomy of plants, arguing that correspondences existed between animals and plants—and even plants and minerals— and that there must be a hierarchy of forms. He believed that plants had glands and some organs, and that their sap was analogous to blood. This led him to the problem of the boundaries, overlaps, and analogies in the three kingdoms of nature. He believed he had found respiration and circulation in plants, and that the ovary of the plant and the uterus of the mammal were the same organ, seeds and fetuses the same kinds of being. He found peristaltic motion in the vessels of both plants and animals, and, though he later repudiated as exaggerated the claim that plants were animals (defended on the basis of a spurious macroscopic comparative anatomy by Ovidio Montalbani, who edited Aldrovandi's *Dendrologia* in 1668), he had identified a distinct realm of organic life in a period in which the boundaries of the three kingdoms were notoriously variable. In an unpublished manuscript he rejected the idea that minerals are alive and grow by interstitial deposition, like plants. They had no organs, he said, no sap flowed in them; the formation of crystals occurred mechanically, as filaments arranged themselves in superimposed planes.[96] Yet he could not avoid noticing that the hardening of bones was in some ways like the freezing of ice, beginning at various nodes and expanding.[97] In the last five years of his life Malpighi was still busy observing and recording the spermatic vessels, the stomach, the anatomy of the bee. Leibniz visited him on his Italian journey in 1689, and they discussed the growth of plants and respiration and looked at sprouting date seeds through the microscope together.[98] The day before Malpighi died, on 29 November 1694, he expressed his worries about finishing a memoir for the Royal Society, fretting that certain things about the anatomy of laurel seeds and the ear of the eagle still needed elucidation.[99]

Malpighi's correspondence testifies to the vigorous pursuit of microanatomy in the Italian medical schools and in Holland. He communicated regularly on the subject with Borelli, who was older, with Lorenzo Bellini, who was younger, and with the Dane Nicholas Steno, who lived in Italy. Of the five classical microscopists, Malpighi's scientific relations were the most satisfactory, though he became increasingly embittered about the reception of his scientific work as he grew in fame and began to draw attention and criticism to himself. Where his specific techniques and his instruments were concerned, he was, however, curiously silent, confirming the impression that self-definition as an anato-

[96] Adelmann, *Marcello Malpighi* 1:477.
[97] Malpighi, letter to Belloni, 6 August 1686, in ibid., p. 503.
[98] Adelmann, *Marcello Malpighi* 1:551–52.
[99] Ibid, p. 658.

mist and observer of nature tended to preclude self-definition as an optician; within this group of observers and recorders, only Hooke—the
least-dedicated observer and recorder—wrote systematically about the
technology of the instrument. But there is no doubt about Malpighi's
dependence on the microscope. After his house burned down in 1668,
he mentioned repeatedly the catastrophic loss of his microscopes, which
left him unable to work. However, he nowhere took the trouble to describe his instruments and their specifications, referring vaguely only to
a flea glass of one lens, which he used to examine the lungs of the frog
and plant galls, and a microscope with two lenses that enabled him to
illuminate the specimen from below through a transparent plate.[100] He
used instruments made by Divini—"piccioli cristalli"—at least from
1671 onward.[101]

To what extent did this division of roles between anatomist and optician contribute to the apparent stall of microscopy after the deaths
of the first generation of microscopists? Was the technically minded
Hooke too distracted an observer, and Leeuwenhoek too convinced of
his role as explorer, for either to ground the subject both technologi-
cally and programmatically? This question is not easy to answer. It is
not known to what extent, for example, Descartes's desire to pursue
microanatomical research was frustrated by his difficulties in dictating
to artisans,[102] and Swammerdam's and Leeuwenhoek's lack of interest in
training successors for themselves by imparting their store of practical
experience and pointing out areas for further investigation gave a certain illusory air of finality to their oeuvre. But Hooke had done exactly
this and yet had failed to establish a program of microanatomical research in England.

Because of the particular constitution and vigor of the Italian medical
schools, the fate of English microscopy was not shared in Italy, where
pathological anatomy developed and became increasingly strong as an
outgrowth of these early studies. Malpighi had provided, in the polemical literature educed by his academic enemies and in the preface to his
De viscerum structura, a splendid defense of microanatomy, in which
he articulated an idea of the body as an ensemble of micromachines
whose true structures and operations are invisible to normal vision.
This was Cartesianism, but with ocular evidence to back it up. Living
processes, he said, are only mixtures and separations: the concoction

---

[100] Malpighi, *De pulmonibus epistola altera*, in ibid. 2:828–29.

[101] Adelmann, *Marcello Malpighi* 2:829.

[102] According to Legrand, Descartes decided six months after its publication that the
Geometrie, and not the *Dioptrique*, furnished the best illustration of his method. He did
not want the practical difficulties of application to reflect negatively on the method; see
"De l'invention des lunettes," pp. 15ff.

and straining of fluids. They are accomplished by figure and shape alone in the glands of the body, by means of very fine pores and vessels whose ultimate fineness we cannot see: these fluids, composed of tiny particles, become visible only when enough of a similar type have come together. The major organs are interconnected through the nerves and vessels, so that the entire body can be affected by a disturbance in one gland; the body is one machine, and the organism a single feeling and moving individual, because of the physical connections between its parts.[103]

Steven Shapin and Simon Schaffer have argued recently that knowledge was a new commodity in the seventeenth century, one that was manufactured rather than found. In their words, "experimental knowledge production rested upon a set of *conventions* for generating matters of fact and for handling their explications."[104] Two conventions that they single out as especially important are the multiple witnessing of individual experiments and their replication. But the new experimental science, they argue, did not require everyone to be an eyewitness; rather, it devised modes of communication intended to secure results and to instill conviction even in those who were not present. These modes had to be distinguished from the old hearsay and reportage, which were as likely to establish vulgar errors as truth, but they also had to be distinguished from conceptual demonstrations. The "virtual witnessing" that could be achieved through illustrations of equipment and experiments served these ends of persuasion, and so did the highly circumstantial scientific narrative, in which the piling up of details, including seemingly irrelevant ones, was intended, as in a forensic context, to win credibility on the grounds that the author was more likely to have done and witnessed these things than to be making them up.[105]

It is not entirely clear from Shapin and Schaffer's account whether "matters of fact" are whatever is generated by these procedures, or whether matters of fact are such that these procedures happen to be the best ones for eliciting them. In both cases, the procedures can be said to be "coercive," though not in the ordinary sense of individualized intellectual despotism, simply because they exclude from the domain of knowledge anything not manufactured or elicited with their help. Shapin and Schaffer's statement that "[l]egitimate knowledge was warranted as objective insofar as it was produced by the collective, and agreed to voluntarily by those who comprised the collective,"[106] does

---

[103] Malpighi, *De viscerum structura*, preface, in Adelmann, *Marcello Malpighi* 1:297.
[104] Shapin and Schaffer, *Leviathan and the Air-Pump*, p. 51.
[105] Ibid., pp. 60ff.
[106] Ibid., p. 78.

not rule out either interpretation. But they have often been taken to be arguing that the objectivity of the new experimental science did not depend on the existence of a truth-bearing relation between the products of science—theories, statements, pictures—and the world, but on an agreement to call the kind of knowledge founded on certain practices "objective," thereby privileging it in relation to intuitive and demonstrative knowledge, or to myth and superstition.

However, such an extreme nominalistic view need not be imputed to them, and indeed their intention appears to be to blur the distinction just noted by a kind of relativizing procedure. Rather than asking what transcendental principle of correspondence ensures that our theoretical statements and experimental procedures are, respectively, true of and paths to reality, we ask what propositions and what procedures have resulted in the body of knowledge that we recognize and designate as public and objective, by contrast with subjective and individualized knowledge. Just as, according to Kant, we cannot seriously debate the question whether our cognitive faculties are adequate to knowledge of the world, so we cannot seriously raise the question whether the procedures of seventeenth-century experimental science were adequate to the achievement of objective knowledge. Given that what they produced is defined as objective knowledge, we need only ascertain what conditions made this possible.

The philosophical advantage of this position is clear. As Socrates pointed out, you cannot know what the truth is before you have got it and recognized it as truth. How then can you, in Baconian or Cartesian fashion, devise a method for arriving at truth before you have identified the class of true statements that your method has as its goal? If you have some independent, premethodical means for arriving at a knowledge of what is true, then your method is redundant. This dilemma forces on us the acknowledgment that the notions of truth and method are to some extent interdependent, but it does not settle the question whether objective knowledge is so merely by definition. Where the realist will argue, like Aristotle, that it is the presence and pressure of reality that makes possible multiple witnessing, replication, adequate and consistent visual representation, and the lengthy description of a real phenomenon, the social constructivist will point out that reality seemed to exert no such irresistible pressure before the introduction of authority-wielding collectives and their preferred ceremonies of induction for new facts, so that social arrangements seem to precede, temporally and logically, the establishment of truth.

One may feel, however, that the issue is not thereby settled in favor of nominalism. What Shapin and Schaffer refer to as conventions appeared accidentally in the course of history in the uncontroversial sense that they need not have appeared. But they were not, for all that, arbitrary. The early members of the Royal Society were expected to abide by these conventions, but it is misleading to suggest that this amounted merely to a special form of etiquette. Multiple witnessing, replication, virtual witnessing, and experimental discourse were all intended to repress distorting features of subjectivity, and to the extent that we can reasonably dispute whether they succeeded in doing so or not, it is apparent that they did not work entirely by convention.

The early deployment of the microscope involves all the features of knowledge constitution mentioned by Shapin and Schaffer and so furnishes excellent material for evaluating their conclusions. Multiple witnessing and replication played a critical role in determining the reactions of the Royal Society to Leeuwenhoek's discoveries; prolixity in description was Leeuwenhoek's literary trademark; and Hooke's iconography went a good way toward establishing the existence of the microworld as fact. Yet I suggest that procedures and institutions are not by themselves constitutive of objective knowledge. It is reasonable to ask whether the supposedly truth-constitutive procedures of seventeenth-century microscopy did in fact produce objective knowledge. The issue is not whether any groups that supported microscopy's practice and development were in a position to produce error-free results, but whether they were confident of their own abilities in this respect, whether they understood themselves to wield the power the conventionalist ascribes to them. And although a group like the Royal Society might wish to consider itself an arbiter of truth and falsity and might specially install procedures in which it had some faith, supporting them with a rhetoric of objectivity, what we mean by knowledge and what they meant was always something other than whatever was turned up by these procedures. The practitioners of protoscience knew as well as we that multiple witnessing and replication did not prevent error, that prolixity in description did not always contribute to shared knowledge but might appear to isolate the writer in a dream world, and that microscopical iconography was fraught with problems.

It might seem that, in advancing that charge, one is simply declaring one's own scientific etiquette to be better at truth production than the etiquette of the seventeenth century, not demonstrating that truth is prior to method. How might one acknowledge the paradox of the *Meno* and still maintain this? This is only possible if one imagines that the facticity of what exists shapes the formation of practice, that it produces the methods used to grasp it in description, a possibility that seems to be ruled out by the localized and contingent character of scientific development as well as the proposal's animistic overtones. But the idea is not, for all that, absurd. Recall the main features of Bacon's method. It is, he says, a way of disturbing and distressing nature, of getting it to stammer out secrets under provocation, for "the nature of things betrays itself more readily under the vexation of art than in its natural freedom."[107] The stress on artificiality is correct, but the image is nevertheless misleading. For nature does not talk, as opponents of correspondence theories of truth never tire of reminding us; it is we who talk, who frame propositions and theories, construct models, and draw representations. The picture of nature as being driven to confession is not, however, an essential component of Bacon's theory of method. Methodicality is a means of eliminating subjective distortion. Beyond that, all methodicality that is successful, including the conventions of seventeenth-century science, is a means, it might be said, of amplifying and focusing the vague pressure constantly exerted on us by what exists, enabling us to form a representation of it. The Paracelsians were right to suggest that our theories are a report of what we have overheard (to translate for a moment into their preferred modality), but wrong to suppose that an extreme degree of openness, attentiveness, and spiritual preparedness was the key. For all that, method cannot take us to what exists; one can agree with the constructivists that it is absurd to say that our methods are a path to objective knowledge, which lies, waiting to be discovered, in the thickets of error. Our theories are, indeed, intellectual constructions. But they constitute, at any given moment of history, our best attempt to say how it is when reality pushes against our instruments and our procedures, instruments and procedures whose reliability for us is determined both by the internal virtues of the theories they help us derive and by the practical use to which the theories can be turned. The result is that method plays a more modest role than either the realist or the social constructivist, who both overstate its role, will allow. It is neither a path to truth nor a set of rules that prescribe what truth is. Method is only a help to the generation of descriptions that would not otherwise be produced, and whose acceptance depends upon factors other than how they were generated.

This brings me finally to a sharper formulation of the question of the relation between scientific instruments and techniques and the notion of objectivity. Earlier I argued that objectivity is less a matter of the selection of a fundamental ontology of a specially colorless, affectless sort than a matter of the observer's relation to the object studied, and of the precautions the observer surrounds himself or herself with to ensure that this relation is not disrupted and that the information obtained from it is maximal. Hooke already knew how to speak the language of

---

[107] Bacon, *The Great Instauration* and *Plan of the Work*, in *Works* 8:48; cf. *New Organon*, in *Works* 8:134.

objectivity, the language of "scrupulous choice, and a strict examination, of the reality, constancy, and certainty of the Particulars,"[108] of the
need for a sincere hand and faithful eye "to examine, and to record, the
things themselves as they appear."[109] The peculiar lesson of seventeenth-
century science is that the direct approach to the knowledge of nature
was even more difficult than the moderns realized. It might seem a kind
of paradox that the route to nature that produces the most adequate
and distinct knowledge is the route that requires the heaviest and most
elaborate interposition of scientific instruments and apparatuses. When
Hooke speaks, for example, of the coordination of hand and eye, one
remembers that hand and eye are *not* coordinated in microscopical
work, that a new series of coordinated movements must be learned,
with some difficulty. And involvement with "things" turns out to involve not only delight and gratification, movement and progress, but
the resistance offered by everything real, and the boredom, frustration,
and even disgust associated with every manifestation of facticity. Like
the brilliant Bichat, who would claim more than a century later that his
books were the sick and dead bodies of men, Boyle crows in the *Christian Virtuoso* that the virtuosi "deservedly cherish the laborious industry of anatomists, in their inquiries into the structure of dead, ghastly,
and oftentimes unhealthfully as well as offensively foetid bodies."[110] The
conversion of minute anatomy and minute natural history into something worthwhile and dignified—on a par with the study of the heavens
and their distant, luminous objects—was the work of apologists and
interpreters who attempted to fit this knowledge into broader contexts
of meaning, with results that could never prove entirely satisfactory.

[108] Hooke, *Micrographia*, preface.
[109] Ibid.
[110] Boyle, *Christian Virtuoso*, preface, in *Works* (1744 ed.), 5:511.

4

Preexistent and Emergent Form

THE MAGNIFYING lens was called into play early in the study of generation, with results that attest both to its role in the formation of positive
science and to its ability to be pressed into the service of myth. In his
defense of the microscope's usefulness, Friedrich Schrader cited in 1681
the display of the "actual" rather than the "potential" preformed plant
lying hidden in the seed—"non potentia . . . sed actu delitescere"—as
one of the instrument's most important revelations.[1] Indeed, the microscope's contribution to the destruction of the form-matter dichotomy,
the bare potentials of scholastic ontology, and the spiritistic agents of
Renaissance nature philosophy suggests that the production of new phenomena is a powerful weapon against metaphysics. Internal criticism,
including the charge of unintelligibility, is capable perhaps only of reflecting and reinforcing, by rhetoric and reasoning, a changed interpretation driven by the phenomena, not of initiating conceptual change.

On the other hand, the seventeenth-century theory of generation has
never been regarded as a display piece for scientific positivism. Between
1670 and 1705, as Jacques Roger has shown, the theory of preformation, in its ovist form, became solidly established in France and was
little threatened by its rivals.[2] Yet its triumph depended less on the quality of observational evidence for it, which was ambiguous and fragmentary, than on metaphysical considerations about order and agency. Not
without reason, the theory in both its ovist and its spermaticist forms is
typically regarded as an embarrassment of early microscopy. It is a favorite example of philosophers who maintain that expectations and theoretical commitments determine the character of sensory data, and who
hope thereby to show that science is not a construction from elementary
observations. Certainly science is not such a construction. But this does
not imply that ideology is primary and that visual experience lags behind as secondary and conditioned. What was seen by the early microscopists, in all its indefiniteness, went into the creation of a picture with
a certain philosophical resonance, and the features of this picture as it
emerged tailored the interpretation of what was seen, leaving behind the

[1] Schrader, *De microscopiorum usu*, p. 16.
[2] Roger, *Les sciences de la vie*, pp. 263ff.

conviction of progress on all fronts in the form of a perfect harmony among reason, theology, and natural science.

There is thus room for skepticism about whether the rise of theories of preformation really illustrates the dependence of perception on prior conceptualization and expectation, whether the story of the theory of generation does not teach us rather how a perceptual limit presents a blank canvas on which the imagination writes what it cannot see, extrapolating from what it can. This is patently the case in Harvey's theory of generation, which is based on a perceptual limit at or near the human optical norm; it is equally true in those later theories that emerged when that limit had been moved back. Indeed, an underlying theme of Roger's great study of early modern biology is that, under the influence of mechanism, biology persisted in evading its subject until the mid–eighteenth century, resting in the imagination. Preformation is, on his view, not a theory of generation at all, but a theory of existence.[3] Although the paradoxical phenomenon of avoidance he describes was pronounced only in France, and only after 1680—in England there were stronger and more persistent currents of antimechanistic thought running from Henry More and Ralph Cudworth to Newton, and less deference to the *esprit de système*—the larger point is unaffected. But evidence for the dominance of ideology should not obscure progress through haphazard empiricism: between the projections William Harvey makes into what he cannot see and the projections made by Marcello Malpighi and Antonio Vallisnieri, there is a world of difference. For Harvey, despite his lucid appeals to experiment and observation, is philosophically committed to the invisibility of a process; his successors recognize only their own limitations in this respect. Premodern theories of generation astonish us not only by their sketchiness but by their fantastic air of completeness, the conviction of their authors that they have said all that it is humanly possible to say on the matter. Unlike the epigenetic theories it replaced, the theory of preformation implied a second-order confidence, a confidence not simply in what future observations could or would reveal, but in the inevitability of making them.

Generation had been one of Aristotle's specialties. In the *History of Animals* he describes the sequence of development in chickens' eggs and the fetuses of marine animals,[4] and in the *Generation of Animals* he gives a philosophical account of how generation can occur. The problem, he thought, was not what the fetus was made of—this was either the egg of the oviparous, or the menstrual blood of the viviparous, fe-

---

[3] Ibid., p. 439.
[4] Aristotle, *History of Animals* 561aff., 565aff.

male—but what made it. And whatever made the fetus had to possess intentionality, to be soul or soul-like, for making is molding, forming, and coloring. "In the early stages," he says, "the parts are all traced out in outline; later on they get their various colours and softnesses and hardnesses, for all the world as if a painter were at work on them, the painter being Nature. Painters, as we know, first sketch in the figure of the animal in outline, after that go on to apply the colour."[5] We cannot suppose that the indefinite form of the embryonic animal is due only to our inability to see its parts; rather, because we cannot see them, we can conclude that the parts are not there from the beginning: "[o]ur senses tell us plainly" that they are not. This argument is less circular than it seems; what Aristotle means is that we see gradual development and the emergence of form, not just growth. We know that the parts of the embryo are there potentially, for otherwise they could not be brought forth at all, but we see that they are not there actually, not in miniature. The molding of the individual organs occurs sequentially, "as Orpheus saw when he compared the formation of an animal to the plaiting of a net," and any unprejudiced observer can see that it is so: "this is certain because although the lung is larger in size than the heart it makes its appearance later in the original process of formation."[6] First the head region is formed, then the upper body, later the lower limbs.

What sets this evocation of actuality out of potentiality in motion is the male semen.[7] Considered as a kind of fluid matter, it appears no more active than the female catamenia, but the semen, or a spirit or nature in it—or it together with heat, air, wind, sun, heaven, and Jupiter—is the proper cause of generation. By analogy one could say that the semen sets the menstrual blood as rennet or fig juice sets milk. It is a case of an active principle working on matter. "There must needs be that which generates and that out of which it generates; . . . if the male is the active partner, the one which initiated the movement, and the female *qua* female is the passive one, surely what she contributes to the semen of the male will not be semen but material."[8]

One may well wonder whether Aristotle's main explanatory categories—passive and active, potential and actual, material cause and formal cause—have not been surreptitiously derived from the phenomenology of the very occurrences they were supposed to explain, and whether their air of obviousness does not derive from this. The idea that the child is sketched and finished like a picture survived in Renaissance

---

[5] Aristotle, *Generation of Animals* 743b.

[6] Ibid., 734a.

[7] Note that Aristotle, *Parts of Animals* 636b, defends a two-semen theory: "she too emits a fertile seed."

[8] Aristotle, *Generation of Animals* 729a.

literature;[9] even Descartes fell under the spell of the analogy of the clay and the potter.[10] But there were rival theories of generation circulating in Aristotle's time that took other physical processes as models. Anaxagoras, who believed that substances could not be generated anew, offered the theory of the *homeomere*, saying that particles of all substances, flesh, bone, hair, and blood, must already exist and come into combination in some way during generation. The observation that earth, sand, and lead scrapings mixed with water will first make a uniform mixture, but, on the addition of air, separate back into their original constituents, like aligning with like, is said by Hippocrates in *De natura pueri* to explain the generation of flesh in all its varieties. If Aristotle's is a theory of epigenesis, these theories are the ancestors of pangenesis, the doctrine that the offspring is formed by a collection of particles drawn from each part of the parental body and reassembled into a complete creature. Both accounts can both be contrasted with the doctrines of preformation and preexistence, which taken together imply the eternality of seeds containing in some way the entire plan of the future animal.

Observations of chicken eggs were continued in the sixteenth century by Fabricius ab Aquapendente and his pupil Volcher Coiter, and, in the seventeenth, by the French anatomists Riolan, and the Englishmen William Harvey, Nathaniel Highmore, and Kenelm Digby. Jean Riolan was using lenses to study aborted embryos as early as 1618, and Nathaniel Highmore was studying chicken eggs with hand lenses in 1651.[11] The earlier set of observations had revealed little that was new: disputes continued over whether the chicken develops from the cicatricula on the yolk or from one of the chalazae, the white, stringy portions on its side. Later observations, however, settled the question in favor of the yolk. It was there that Harvey saw, at the end of the third day of development, "a capering bloody point, which is yet so exceeding small, that in its *Diastole*, or Dilation, it flasheth only like the most obscure and almost indiscernable spark of fire; and presently upon its

---

[9] Thus Lemnius says that "we may observe the same in painters, who first with a more rude pencil, or with a cole [sic] or chalk draw a picture in the ground work of it . . . then they polish it and finish it so that those things that before appeared rough, hid, undressed, dark, obscure, shadowed, do afterwards show neat, pleasant, and clear." *Secret Miracles of Nature*, p. 302.

[10] "I shall suppose the body to be nothing but a statue or machine made of earth, which God forms." Descartes, *Treatise of Man*, in *Philosophical Writings* 1:99.

[11] The painting of Harvey with a compound microscope is thought by his biographer to be fanciful; see Keynes, *Life of William Harvey*, p. 341 n. 1. There is a reference to "perspectives" nevertheless in Harvey, *Anatomical Exercises*, p. 93, and to observations on the hearts of wasps, hornets, fleas, and lice, and one to magnifying-glasses in the *Disputations*, p. 96, which are said to be as good an aid to vision as bright sunlight.

Systole, or Contraction, it is too subtile for the eye, and quite disappeareth."[12]

His own examinations convinced Harvey that what Fabricius had said was outrageous: his predecessor, he charged, "layeth aside the verdict of sense, which is founded upon dissection; he flies to petty reasonings borrowed from mechanicks" when he describes the fetus as though it were a house or a ship whose frame or skeleton is laid down first.[13] Nature does not lay down the hard parts first and the softer parts over them, as though it were a mechanic. It is rather the heart and the blood, where the soul resides, that are there at the beginning and build for themselves a future dwelling place. "Applying the ribs and sternum as a defence," Harvey says, the embryo "walls itself about."[14] At the end of the fifth day, the soul is embodied. The head and eyes of the fetus appear, and it continues to grow and develop, "just as mold grows elsewhere in damp places between the chinks of walls of houses which lie long unswept or like camphor on planks of cedar wood and lichen on stone and the bark of trees, or lastly, as fine down is bred on caterpillars," until the egg hatches on the twenty-first day.[15] This process is the same for all the more perfect animals. In the early stages of development, dog, horse, deer, ox, snake, and human, Harvey says, "so exactly resemble the shape and consistency of a maggot that your eyes cannot distinguish between them."[16]

In this superior form of generation, the animal is both creator and created. It "attracts, prepares, elaborates," and makes use of the material out of which it builds itself; the processes of growth and formation are simultaneous. Only in insects does "metamorphosis," in which decaying matter is acted on by a plastic force that transforms it into a creature, occur. On the accessory causes of generation, Harvey is as circumstantial as Aristotle; one must take into account not only the soul in the heart that builds the body but incubation and temperature, the position of the sun in the zodiac, and even Fabricius's generative and architectonic faculties: immutative, concoctive, formative, and augmentative. There is something divine, spiritous, and operative, "analogous to the essence of the stars," resident in the sperm that sets the process in motion.[17]

Though he might seem bound by the Aristotelian tradition, Harvey saw himself as having broken decisively with the ancients and with his

[12] Harvey, *Disputations,* p. 96.
[13] Ibid., p. xlix.
[14] Ibid., p. 341.
[15] Ibid., p. 111.
[16] Ibid., p. 113.
[17] Ibid., pp. 183, 343.

immediate predecessors alike.[18] "Everything that has been handed down
to us from all antiquity concerning the generation of animals is erro-
neous," he says. The opinions of Aristotle are to be "refuted" and "re-
buked." Why he understands his own version of epigenesis to be very
different from Aristotle's will become clear; but note here the familiar
early modern emphasis on seeing for yourself, on not acquiring your
knowledge through the discourse or even the pictures of other investigators. It is "unsafe and degenerate," Harvey says,

> to be tutored by other mens commentaries without making tryal of the things
> themselves; especially since Natures Book is so open and legible. . . . what we
> discover by the senses is more clear and manifest to us than that which we
> discover by the mind, because the latter springs from these sensible percep-
> tions and is illuminated by them. . . . Wherefore it is that without the right
> verdict of the senses controlled by frequent observations and valid experience,
> we make our judgements entirely on phantoms of apparition inhabiting our
> minds. . . . We must, I say, rely on our experiences and not on those of other
> men. . . . Therefore, gentle Reader, take on trust nothing that I say on the
> generation of animals. I call upon your eyes to be my witnesses and my
> judges.[19]

Harvey seems here to recommend his own book partly for what it
lacks—the illustrations supplied by Fabricius—and he calls on the
reader not to believe him, or rather only to believe him insofar as the
reader is ready to reproduce his observations. This points to a set of
fundamental difficulties in the establishment of early modern science:
the problem of establishing trust in the medium discredited by one's
predecessors, of giving one's own books a special relationship to the
Book of Nature that theirs did not have. Harvey and Descartes were
both innovators whose researches were conducted in solitude: they were
not academicians who presented themselves as without individual
talents but who posited their group, with its proving and re-proving, as
a locus of authority by consensus. Though they appealed to their
readers—Descartes by subtle indirection—to imitate them, they made
no attempt to divest themselves of personal authority or to disperse this
authority in a community of eyewitnesses. They tried instead to establish themselves as singularly trustworthy observers, citing their freedom
from tradition, their habits of clarity and exactness. The plea for the
recognition of exact individual observership or the application of strict
individual methods of reasoning is nevertheless addressed to a community that must be at least somewhat receptive to these values to start, a

---

[18] On Harvey's epistemology vis-à-vis that of Aristotelians and contemporary anatomists, see Wear, "William Harvey."

[19] Harvey, *Disputations*, pp. 12–13.

reminder that representations of the relationship between the individual and the group tend to be simpler than the reality.

Harvey's insistence on the primacy of immediate, first-person sense experience should not delude us about the role played in his theories of fertility and conception by *invisibilia*, by what he did not see, and what he thought no one would ever be able to see. He was the last of the great macroscopic anatomists to work on the theory of generation, and his theory may be regarded as an attempt to come to terms with the invisibility of that process, within the limits of a program of exact observation that, despite its brave declarations, found itself obstructed and deflected. The *Disputations Touching the Generation of Animals* was published late in Harvey's life, in 1651; it probably represents investigations undertaken over a period of twenty years or more. It was published in Latin, then translated into English almost immediately, and Latin editions were published in Holland and Italy in 1651, 1662, 1666, and 1680, giving the book a wide circulation.[20] The famous frontispiece with its motto *Ex ovo omnia* shows Jove opening an egg out of which a variety of creatures—reptiles, insects, a quadruped, birds, and a human being—tumble. This legend has been the subject of much controversy: Harvey did not mean by "egg" anything so exact as the product of a parent animal capable of producing a creature similar to it. He seems rather to have regarded the egg as any vital primordium, which might take the form of a seed, an egg, or even a worm, and which might or might not have had a parental creator similar to the offspring that would emerge from it.[21]

The point of the motto is nevertheless that there is one kind of generation that occurs in both the oviparous and the viviparous. As Harvey's friend Dr. Martin Lluelyn put it, there is no doubt

That both the Hen and Housewife are so matcht,
That her Son Born, is only her Son Hatcht,
That when her Teeming hopes have prosp'rous bin,
Yet to Conceive, is but to lay within.[22]

Aristotle's catamenia is irrelevant to generation: it is present in quantity only in women and not in other females anyway. It is true that we do

[20] Keynes, *Life of William Harvey*, p. 360.

[21] Harvey is thought to have written a lost or burned treatise on the generation of insects; his plans to write a new treatise refer to animals "generated by Metamorphosis, namely . . . Insects, and Spontaneous Productions, in whose Egges, or first Rudiments there is a plaine Species or Immaterial forme." Ibid., p. 358. He seems to have allowed that creatures could be generated by parents of a different species; see Harvey, *Anatomical Exercises*, p. 3.

[22] Prefixed to the first English edition (1651) of Harvey, *Disputations*.

not find anything in the uterus of the vivipara that resembles the unfertilized egg that can be found in the hen's ovary. But the egg is there, formed by the uterus after fertilization: perhaps it is to be identified with the fetus itself enclosed in its amniotic membranes. Harvey did not know.

Harvey rejected Aristotle's notion that females are passive and supply only the matter of generation, pointing out that the female hen must possess a formative power because she can make eggs, even if these are sterile without the male. But both the male and the female are, he says, only "instrumental efficients." Both derive their fertility from a higher power of nature, the sun or God. It is as though the female received the "art, concept, form and laws of the future fetus by copulation with the male."[23] But this is no cause for inordinate pride in males, for "[t]he cock does not confer any fecundity on the hen or on the eggs by the simple emission of his geniture, but only insofar as that geniture is prolific and imbued with formative power, that is to say is spiritous, operative, and akin to the element of the stars."[24]

The failure to find a visible mammalian egg was the outcome of the studies of vivipara that Harvey undertook between 1630 and 1635. Opening daily the bodies of mated does furnished him from King Charles's park, in imitation of his egg experiments, Harvey found nothing for two months. "I dissected diverse of them, and discovered no seed at all residing in their Uterus; and yet those whom I dissected not, did conceive by virtue of their former Coition (as by Contagion) and did Fawn at their appointed time."[25] He also found no trace of male semen after coition, which, he tells us, angered the king's physicians, who "persisted stiffly that it could no waies be, that a conception should go forward unless the males seed did remain in the womb."[26]

Most doctors did in fact subscribe to the Galenic theory that male and female semen are blended in the womb and initiate conception, a theory that explained the presence of similar organs in the male and female, the female's "emission," and the evident dual heredity of the offspring.[27] Harvey did not like this any better and waxed indignant

[23] Harvey, *Disputations*, p. 154.

[24] Ibid., p. 183. Here Harvey appears to be following Renaissance theorists who discuss the operation of "plastic principles." See Pagel, *New Light on William Harvey*, pp. 100ff. Jacob Schegk (or Degan) (1511–87) wrote on the formative virtue of the semen, departing from Aristotle by treating it as an active principle that was neither animate nor inanimate, neither an individual soul nor matter.

[25] Harvey, *Anatomical Exercises*, pp. 416–17.

[26] Keynes points out that deer are peculiar in this way; their embryos are long and narrow and difficult to identify. *Life of William Harvey*, p. 346.

[27] On two-semen theories, see Roger, *Les sciences de la vie*, pp. 271–72. A problem with the theory (and with ovist theories as well) was the distance between the female testicle and the fallopian tube.

over the "so-called" female testicles.[28] In 1661, Antoine Everaerts confirmed suspicions of their relevance by showing that extirpation of the
female rabbit's "testicles" induced sterility; this result was not known
to Harvey, who died in 1657. He did not recognize the mammalian
ovary as such, and his opposition to two-semen theories prevented him
from attaching significance to the organ itself. Because he could not
deny that the hen, the doe, and the woman are essentially the same in
their reproductive structure, he could find a way out of this impasse
only by declaring the relation between visible, tangible egg and visible,
tangible sperm to be irrelevant. We cannot see what in the nature and
action of the semen makes the egg fertile, so it is not so surprising after
all that we cannot always see the egg at the start. The philosophy of
seeing for yourself issues in a declaration of the absolute invisibility of
the process of generation.

Harvey's main target here is what he calls "vain, mechanical explanations"—appeals to processes—such as curdling, or what he sees in Aristotle's suggestion that the semen touches the menstrual blood or the egg
of the fowl and thereby sets something in motion as in an automatic
machine.[29] In conception there is no mechanical action; there is not even
any touching: "A woman after spermatic contact is made fecund by no
perceptible corporeal agent, and is affected in the same way in which
iron touched by a loadstone is immediately endowed with the virtue of
the loadstone and draws other iron bodies to itself."[30] Something passes
from one to another, but this something is not corporeal, nor is it something that simply makes the egg develop into a chicken. Fertility acts on
the uterus and ovary, for a hen that is fertilized once may go on to
produce numerous fertile eggs after the male has been removed. It is
thus fertility itself—a power—that is transmitted, for it is not only a
chicken that comes to be, but a chicken possessing the same reproductive power as its parents.[31]

Harvey allows that this power that is transmitted is something, only
not something corporeal; it resembles an effluvium, an odor, or a ferment, none of which, being untouched by Cartesian dualism, he regards
as material. "What makes the seed fertile," he says,

is on the analogy of an infection . . . the generative seed, just as it passes from
the male, lies dormant in the woman as in warm fermenting matter. . . . Or
else like a . . . light in stone . . . the pupil in the eye, in sense motion . . . in the

---

[28] Harvey, *Disputations*, p. 343.
[29] Aristotle, *Generation of Animals* 734b.
[30] Harvey, "On Conception," in *Disputations*, p. 443.
[31] Harvey, *Disputations*, p. 157.

body. . . . Like a ferment, vapour, odour, rottenness . . . by rule. . . . Or like the smell given off by flowers. . . . Or when a soul is a god present in nature, that is divine, which it brings about without an organic body by means of a law.[32]

These extraordinary examples of parataxis from his notebooks return us again to Renaissance nature philosophy and the volatile effluvia that are the real essences of a substance, the medicines that are efficacious in virtue of their odors. The analogy between contagion and conception is particularly striking in this context. In both, Harvey says, "there is inherent . . . some kind of life and divine principle by which they grow of themselves and propagate diseases like to themselves as a result of contagiousness in the body of another person."[33] "In like manner as Physitians observe, . . . contagious diseases (as the Leprosie, the Pox, the Plague and Pthisick) do propagate their infection and beget themselves in bodies yet sound and untoucht, meerly by an extrinsecal contact; nay sometimes only by the breath and . . . by inquination; and that at a distance, through an inanimate medium, and that medium no way sensibly altered."[34] The intimation that the propagation of diseases was in some way like the production and proliferation of offspring, that a disease was a kind of alien life that might grow and decline, was felt as such before any precision could be given to the idea at all, and long before the pieces of the puzzle were gradually assembled. But as if all this were not curious enough, Harvey went on to draw a parallel between the brain and the womb. Both generate conceptions: imaginations and thoughts in one case, a fetus in the other; and the surface of one is visually similar to the surface of the other, both being soft, full, and smooth.[35] He meditated further on this theme: the mind forms a conception of something absent and external, a desired object, and from this conception desire for the thing is increased. And so with the child; the womb's desire for something absent and external produces a conception, which enhances the desire and makes it formative.[36] There is little here of the matter-of-fact tone of Aristotle's essays on reproduction: with Harvey, the barnyard becomes a magic realm. His patient hours of watching revealed to him the anthropomorphic qualities of fowls in the pride, ornament, and aggression of the roosters driven by life forces. And he came to see human behavior in animal terms; the wit, beauty, and pride of young people in their courtships is simply

[32] Harvey's manuscript notes, as given in Whitteridge's appendix to ibid., pp. 455ff.
[33] Harvey, *Disputations*, p. 147.
[34] Harvey, *Anatomical Exercises*, p. 254.
[35] Harvey, *Disputations*, p. 451.
[36] Ibid., pp. 452–53.

nature working through them, as it works through every creature, man, woman, and chicken. When parents have completed their work they wither and grow old "and as if God and Nature had forsaken them, they decline speedily, and hasten to their end, like creatures weary of their lives."[37]

The theory of preformation might seem to depend on a mistaken visual and literary recording of what was seen, an illusory *micrographia*, a projection of belief into sensory material, where the theory of epigenesis from Aristotle to Harvey was by contrast the faithful reading of direct experience. But the relation of epigenesis to experience is complex. How did this faithful record come to seem a paradigm of superstition and obscurity? As Daniel Fouke has remarked, in rejecting scholastic forms and virtues and Renaissance plastic natures, the moderns equated the truth of an explanation with its intelligibility, and its intelligibility with its visualizability, or with the analogical similarity of the process to a visualizable process.[38] One might add that they regarded intelligible explanations as those that provided for the possibility of mechanical simulation: epigenesis, like the action of the magnet, is readily visualizable but cannot be easily modeled in terms of mechanical micro- or macroprocesses. Preformation was visualizable, even if it was never actually observed, and, by treating generation as growth, it made mechanical modeling a possibility. But why should the remote possibility of mechanical modeling have seemed the very touchstone of truth, when before one's eyes was a picture of nonmechanical formation?

Certainly before preformation theory came into its own, philosophers of a materialist or atomistic bent were trying to solve the problem of generation without appeal to Aristotelian potentials or form and matter. These efforts—the accounts of mechanical assembly given by Descartes and by Kenelm Digby and the seed theory of Gassendi—are vague and confusing, and strike the reader as either presumptuous or timid. Writing to Mersenne in 1639, Descartes promised a mechanical theory. "I find [in animals] nothing which I don't think myself able to explain the formation of in detail though natural causes, just as I explained in my *Meteors* that of a grain of salt or a little snowflake."[39] In his treatise on the formation of the fetus posthumously published as an extra chapter of his *Description of the Human Body* in 1648, he makes this claim even bolder. "If one knew well enough," he says, "what all the parts of the semen of some species of animal, for example, *man* are,

[37] Harvey, *Anatomical Exercises*, p. 158; *Disputations*, p. 151.
[38] Fouke, "Mechanical and 'Organical' Models," p. 365.
[39] Descartes, letter to Mersenne, 20 February 1639, in *Oeuvres* 2:525.

one could deduce from this alone by reasons completely mathematical and certain the entire shape and conformation of each of its members, just as, reciprocally, if one knew the particularities of this conformation one could deduce what semen is."[40] But he must apologize for not having had the opportunity to make observations and for being forced to speak in general terms. He bases his account up to here on the two-fluids theory: the male and female seminal liquors meet and mingle and produce a reaction like the frothing of beer. Expansion and pressure, heat and fermentation, furnish the modus operandi; however, there is no serious attempt to account for the formation of organs or their order. "Some of the particles," Descartes says, "assemble towards a part of the space which contains them where they expand and press the other particles which surround them." This cluster of particles begins to form into the heart and to beat.[41] The brain and backbone follow, and then, as new passageways are made, and as pressures build up, the rest of the bodily parts and sensory organs are formed.

Similarly, Digby, who did open some artificially incubated eggs, hastens in his account rapidly over the earliest period of formation with a good deal of hand-waving. The spot of blood at the cicatricula, he tells us, becomes a heart, veins appear around it, and soon "in processe of time, that body incloseth the heart within it. . . . After which this little creature soone filleth the shell, by converting into severall parts of itselfe all the substance of the egge."[42] This is slapdash compared with what Harvey had to say. But Digby does try to supply a philosophical account that relates plant and animal generation, as Harvey had not done, and one original feature of his account is that the formation of the fetus depends crucially on the action of the environment.

He begins by criticizing the old theory of pangenesis, according to which, in the mature organism, superfluous nutriment that would otherwise cause it to continue to grow assembles from every part of the body—from blood, bone, tooth, liver, hair, and so forth—and aggregates to form all or part of a new organism with both the acquired and the familial traits of the parents. This is impossible, he thinks; the individual collection in the body of all and only those particles necessary for the production of a single offspring makes the imagination boggle. Equally important, the theory fails to explain the generation of creatures that do not have similar parents. "[H]ow could vermine breed out of living bodies, or out of corruption? How could ratts come to fill shippes, into which never any were brought? How could froggs be in-

[40] Descartes, *Description du corps humain*, in *Oeuvres* 11:277.
[41] Descartes, *Oeuvres* 2:254.
[42] Digby, "Of Bodies," in *Two Treatises*, pp. 220–21.

gendered in the ayre? Eeles of deewey turfes? or of mudde? Toades of
duckes? Fish, of hernes? And the like. To the same purpose: when one
species or kind of animal is changed into an other: as when a caterpillar
or silkeworme becomes a flye, it is manifest that there can be no such
precedent collection of partes."[43]

One needs to admit the possibility of a mechanical transmutation of
one substance into another, the conversion of blood or water, for example, by pressure into plant or animal substance, rather than countenanc-
ing qualitatively different *homeomere* that are separated from the nutritive mixture. Here the details necessarily become sketchy; Digby is
mainly concerned to show that the action of the environment on the
developing subject is part of the "ordinary and generall course of nature," replacing any divine guidance or the action of a *vis formatrix*. In
a white bean that is germinating, for example, the water inside it causes
it to swell as it is warmed in the earth; the pressure of the earth causes
a compression and a push upward; the sun causes further expansion,
clogging, and finally the expulsion of plant parts. The initially homogeneous seed is thus differentiated by external agents and accidents and by
"steames of circumambient matter." The process occurs as follows:

> Take a beane, or any other seede, and putt it into the earth, and lett water fall
> upon it; can it then choose but that the beane must swell? The beane swelling
> can it then choose but breake the skinne? The skinne broken, cann it choose
> (by reason of the heate that is in it) but push out more matter, and do that
> action which we may call germinating? Can these germes choose but pierce
> the earth in small stringes, as they are able to make their way? Can these
> stringes choose but be hardened, by the compression of the earth . . .?[44]

And so on. He concludes by asking why, "if all this orderly succession
of mutations be necessarily made in a beane, by force of sundry circumstances and externall accidents; . . . may it not be conceived that the like
is also done in sensible creatures; but in a more perfect manner?" And
he rejects the supposition that the parts are already there:

> Surely the progresse we have sett downe is much more reasonable, than to
> conceive that in the meale of the beane are contained in little, severall similar
> substances; as, of a roote, of a leafe, a stalke, a flower, a codde, fruite, and
> the rest; and that every one of these, being from the first still the same that
> they shall be afterwardes, so but sucke in, more moisture from the earth, to
> swell and enlarge themselves in quantity. Or, that in the seede of the male,
> there is already in act, the substance of flesh, of bone, of sinew, of veines, and

---

[43] Digby, *Two Treatises*, p. 215.
[44] Ibid., p. 217.

the rest of those severall similar partes . . . and that they are but extended to their due magnitude, by the humidity drawne from the atmosphere?[45]

So too in animals there is no preformation in the seed, "which containeth not in it, any figure of the animal from which it is refined, or of the animal into which it hath a capacity to be turned."[46] There is only "one homogeneall substance" made of blood. As in Descartes, the entire account is phrased in terms of pressure and blockage, or drying and expanding, density and rarity, and particulate assembly is still the means by which organs are compounded and shaped.

Digby's rejection of the idea that there is a figure of the animal in the seed, or that the parts of a plant exist in miniature in its "meale" or pollen, points to the availability of such accounts. Nathaniel Highmore published in 1651 the *History of Generation,* in which he provided drawings of germinating beans and seeds showing leaves and stalks "folded up," "inclosed," or "wrapped up" in husks and coverings; these drawings were published on the same plate as Highmore's drawing of a chick embryo.[47] Though Highmore believed that the fetus was formed pangenetically from atoms of both parents, such drawings fixed themselves in the mind, as did the tropes of literary culture. In Sir Thomas Browne's exquisite *Religio medici,* the themes of intelligible, nonvisible delineation and divine craftsmanship are integrated with the ocular evidence, as far as it goes: "In the seed of a Plant, to the eyes of God, and to the understanding of man, there exists, though in an invisible way, the perfect leaves, flowers and fruits thereof: (for things that are in *posse* to the sense, are actually existent to the understanding)."[48] Henry Power, his disciple, carried this a step farther, writing to his mentor in 1659 that with lenses one need not simply speculate but can indeed discover that the entire plant is perfectly "epitomized" in the seed, which contains not only its potential form but the plant itself, "capsulated & straddled up in severall filmes, huskes & shells."[49] The smallest seeds are "nothing but their own Plants shrunk into an Atome." "Some say," he continues, "that in the Cicatricula . . . by a good microscope you may see all the parts of the chick delineated before incubation." Power himself did not claim this: in his observations of 1664 he noted only that on the second day of incubation he could see a white spot in the cicatricula "which in futurity proves the Heart with its Veins and arteries." Yet he concluded more than he saw. "So admirable is every Organ of this Machine of

[45] Ibid., p. 219.

[46] Ibid., p. 223.

[47] Gunther, *Early Science in Oxford* 3:206–7.

[48] Browne, *Religio Medici,* in *Selected Writings,* p. 58.

[49] Power, letter to Sir Thomas Browne, 10 May 1659, in Browne, *Works* 4:266–67.

ours framed, that every part within us is intirely made, when the whole Organ seems too little to have any parts at all."[50]

Following Jacques Roger,[51] we might distinguish between the doctrine of *preformation* (the embryo with all its parts is present in a germ in one or the other parent) and the doctrine of *preexistence* (all existing germs were created at the beginning of the world), while noting that the two converge in the doctrine of *emboîtement*. The locus classicus, for Christian philosophers, of the doctrine of preexistence was the writing of Saint Augustine, whose words show that the horizons of perception and the horizons of existence of medieval writers are not, as is often alleged, the same.

> Certain seeds of all the things which are generated in a corporeal and visible fashion lie hidden in the corporeal elements of this world. And these seeds, which are now visible to our eyes from their fruits and living things, are distinct from those hidden seeds of those former seeds. From them, at the Creator's demand, the water brought forth the first swimming things and the winged fowl, and the earth brought forth the first buds according to their kinds, and the first animals according to their kinds. . . .[52] Note that the smallest shoot is a seed; for when properly planted in the earth it brings forth a tree. But there is a more subtle seed of this shoot in some grain of the same species, and up to this point it is visible even to us. Furthermore, there is also a seed of this grain; although we cannot see it with our eyes, yet by our reason we can conjecture that it exists.[53]

Only by assuming the existence of such seeds, all of them present at the creation, but most still lying dormant, can we explain the abundance of animal and vegetable life and the proliferation of creatures that occurs even without sexual commingling.

The theoretician who considered the question of generation most exhaustively at midcentury and who produced the most thoughtful and elaborate accounts was also the one who confessed that the whole subject was a sort of mystery that all his experience, learning, and theory were unable to penetrate: Gassendi. Preexisting, though not fully preformed, seeds figure in his account—the sum total of his writings on generation can hardly be called a theory. As always, his authority is Epicurus, who also thought that seeds came into being at the first formation of the world through the mixing and combination of atoms. Gassendi softens this dangerously pagan doctrine by noting that God

---

[50] Power, *Experimental Philosophy*, p. 60.

[51] Roger, *Les sciences de la vie*, p. 326.

[52] Classical theory also recognized two kinds of seeds, visible and invisible, the latter resident in air and water. See Varro, *On Agriculture* 1.40.1.

[53] Augustine, *Trinity*, book 3, chap. 8, pp. 108–9.

might have created all seeds at once, but holds it possible—one senses that this is the preferred account—that he created only atoms or corpuscles with mass, shape, and motion, whose collisions, minglings, combinations, and interweavings now form "molecules or small structures similar to molecules, from which actual seeds are constructed and fashioned within the plant."[54]

Gassendi was a sort of materialist, celebrated for his attacks on Cartesian dualism, but in view of his complicated, image- and intention-driven accounts of generation and his distribution of material souls throughout nature, he cannot be described as a consistent mechanist. His seeds are alive and ensouled. The soul of the plant is corporeal, though its substance is highly refined and diffused through the whole plant in which it resides; "a substance which, like spirit or a flammule, is singularly thin, pure, highly active, and assiduous, all of whose parts communicate with one another in such a way that in whatever part of the plant it is, it contains the idea, so to speak, and impression of the other parts" (p. 799). The soul of the seed is an offspring of the soul of the plant, an *animalula*, an "epitome," and it dutifully copies its parent. "As if it had been taught and was serving its apprenticeship," it has "already begun to do whatever the whole soul knew how to do throughout the plant" (p. 800). What the greater soul was doing in the plant in animating differently the diverse parts—roots, stems, leaves, and so forth—the *animalula* now begins to do in the seed: "the enclosed *animalula* sets about separating all the particles of seed, dividing their regions so to speak, and, as it were, directing the work of each; and in addition, the particles themselves by a sort of will of their own remove themselves from the confusion, like approaching like and uniting with it" (ibid.). Although the parts of the plant-to-be are not all present from the beginning, the movement and combination of corpuscles produces their rudiments early on, and they then gradually "unfold":

> While the bud is sprouting, it becomes apparent that the stamina, so to speak, of the whole shoot have already been formed in it, and since these are conglobated and fine, they need only to unfold and grow larger in order to be more distinctly visible. Hence upon inspection the structure of the shoot may be recognized with its pith, wood, bark, also its little leaves, and in their axillae indications of the buds that will emerge in the same or in the following year. . . . This is visible in bulbs, too, and indeed even more clearly if you dissect them when they are beginning to germinate. (p. 801)

Animal seeds, in accord with the two versions of the general theory, were either distributed by God at the Creation when he conferred fe-

---

[54] Gassendi, *De plantis*, in Adelmann, *Marcello Malpighi* 2:798. Subsequent references to this work are cited parenthetically in the text.

cundity upon the earth with his command to be fruitful and multiply, or are even now being formed from the perpetual motion of atoms, "since it is established from what I have said elsewhere that all things are stirred up by continual, internal, quivering movements and that even when things appear to be most at rest their internal principles are nevertheless perpetually struggling, unfolding themselves, revolving and mixing" (p. 803). In case this sounds too much like tychism, Gassendi incorporates divine regulation. The *animalula* elaborates and perfects its organs necessarily, yet not in a blindly mechanistic way, because "an Artificer so great, so wise, and so powerful . . . has produced it as it is, has endowed it with such force, and has willed it to be enclosed in such a body of such construction that it could do no otherwise than act in such a way and build such a structure" (p. 804). Despite these precautions, the old pagan randomness persists, even in an intermediate account according to which the originally irrational and blind concourse of atoms becomes systematic as nature matures and learns to settle down: "Nature gradually grew accustomed to this, learned thus to bring about the propagation of animals similar in kind, and as a result acquired from this perpetual motion and ordering of atoms a certain compulsion always to operate in this way" (p. 808). Where the scholastics described nature as a habit of God's, Gassendi invokes the habits of nature.

In correspondence to Aristotle's distinction between univocal and equivocal generation, Gassendi recognizes both a kind of generation requiring copulation, warmth, and incubation "in a similar animal" and a kind that takes place in putrid material dissimilar to the animal produced. But even the second type of generation involves seeds: there is no generation that depends purely on accidental external causes. Even in the seemingly "spontaneous" production of insects, "every single one of these animals . . . must arise from its own special seed; where we deduce, as if from their footprints, whence these seeds can be drawn, and why they can be found in a particular place and fostered by a particular agent" (p. 805).

Midway through his discussion, Gassendi confesses that, despite the learned parade of the concepts he has made use of—atomic interactions, operative ideas, internal forces, and the habits of nature—nothing has really been explained. "All our ingenuity and fine talk," he admits, "have done nothing to produce a real knowledge of that internal, hidden governance" (p. 806). As a man born in a forest and shown a clock enclosed in the setting of a ring

would only wonder at the delicacy and elegance of its structure, and the long duration, regularity, and spontaneity of its motions and would never guess how the little machine could be made so perfect, so too do our powers fail completely when we are confronted by these achievements of Nature, at the elaboration of which we were present neither as spectators nor participants, and like untutored woodsmen we can only be struck with wonder but cannot divine or conceive by what artifice they have been accomplished; for indeed, each one of them is a little machine within which are enclosed in a way impossible to comprehend almost innumerable [other] little machines, each with its own little motions. (ibid.)

Yet he goes on in the face of this helplessness, trying to fill in the story, making no effort to cleave consistently either to a theory of preexistence and miniaturization or to one of formative action. In fact, he does not seem to see these as exclusive alternatives. The soul contained in the semen of both parents begins to work on the mixture forming a little body, a *corpusculum*. This *corpusculum* is like the image of your face you might see in a small convex mirror: complete in all its parts but reduced in size (pp. 810–11). All the parts of the fetus, he says, seem "to commence at the same time, so that right from the beginning the rough stamina of everything are present together; yet it does not turn out that all the parts are perfected together, or at the same time, but some appear larger and more distinct earlier and others later" (p. 811). Nature does not, as Aristotle had maintained, imitate the artist who must work sequentially, for it, or God, who is present everywhere, can act on all the parts at once. And this it must do because of the interdependence of organs, and the correspondence of the veins and arteries of the heart and liver. This a priori argument is coupled with what would become a familiar transperceptual assertion. Even in chickens on the fourth day of incubation, when the wings and legs cannot be seen, "the beginnings of wings and legs, though extremely small, are nevertheless present at that time with all their articulations and are destined immediately to increase greatly in length, that is to say, they are as if folded up, in the same way in which . . . a branch is folded within the bud and a flower within the bulb" (p. 812). This evocation of sleeping life is continued in a beautiful passage in which Gassendi imagines that even the grain of wheat lying in the granary is not without a soul; it is merely asleep, like the unincubated egg, and although such souls may apparently be lost by boiling, roasting, or crushing, they may survive these ordeals and "be preserved like trees that are dormant through the winter and those swallows of northern climates, and in particular of Muscovy, that grow stiff in the ice," and awaken to warmth and spring (p. 814).

Gassendi's account thus has three phases; it combines mechanism and chance with faculty theory and with preformation. The concourse of atoms evolving over time into a habit of nature, a pattern and plan for reproduction, does not exclude and even seems to require the systematic operation of a corporeal *animalula* acting plastically on the semen or material of the seed. A preformed, though largely undelineated, organism appears to be an immediate product of this action. Though informed by "modern" ideas—corpuscularianism, preformation—Gassendi's discourse can hardly be supposed to give a theory rather than constituting a set of imaginings on the theme of reproduction. And indeed preformationism as just such an imagining displayed itself before the deployment of the microscope in those literary passages that took their force and vividness from three sources: first, the image of God as agriculturalist, fructifying the universe by scattering its surface with tiny, invisible seeds; second, the familiar experience of peeling back a bud in early spring or opening a bean in the process of germination and finding inside it tightly folded leaves or flowers; and third, the image of the book digest or compendium that contains the whole intellectual substance of the original book but in compact form. These images have an economy and coherence that contrast with Harvey's gestural reference to spiritual agencies and contagions, and with the labored but vague discursiveness of the mechanical-necessitarian accounts of Descartes and Digby.

It would be correct, then, to say that a philosophical place had long been prepared for the doctrine of *emboîtement*—the doctrine that the whole organism has always been present, with every part precisely arranged and connected, in the body of the parent—as it emerged in the earliest period of microscopy. But it is not exactly right to say that expectations determined the quality of the visual experience of the first generation of microscopists. Their reportage concerning what they saw—and this includes the illustrations they provided as a record of their experiences—was accurate and, for the most part, unimpeachable. Their interpretations nevertheless exceeded and even contradicted it. The problem for the philosophy of science that is generated is not what we are to do when our theories affect the very character of our observations, leaving us with no certain foundation on which to build. It is rather this: what are we to do when there are strong extraobservational reasons, or observational reasons drawn from outside the immediate context, for supposing that our experimentally generated experience—deep as it is—nevertheless contradicts or is only a partial guide to reality? How are we to decide whether a theory to which we are drawn is really consistent or inconsistent with any given observation?

Examine first in this connection the embryological work of Malpighi. Malpighi's apologists have been eager to rescue him from what they regard as an invidious association with the doctrine of preformation.

Yet it must be admitted that he supplied phrases, ideas, and the authority of his observations to such philosophical preformationists as Malebranche and Leibniz, who were the agents responsible for disseminating and popularizing the doctrine. Even if his position can in no way be construed as implying *emboîtement*, it was possible for philosophers to cull from his texts the motifs that would best serve them.

It is unclear whether Malpighi really believed that the rudiments of an organism might exist in the female egg before fertilization, but he certainly believed that the parts existed before incubation, which is an altogether different matter. He says that in the unfertilized egg he saw only "a globular, white, or cinereous body which . . . when torn open revealed no structure peculiar to and different from it,"[55] suggesting that the material of the egg is not in any way predifferentiated. In any case, his main descriptive treatise, *Dissertatio epistolica de formatione pulli in ovo*, is sequentialist. In its celebrated opening passage, Malpighi confesses that he cannot find a beginning to development: he observes philosophically that the process of coming into being cannot be studied, for one cannot study what does not yet exist or what has already come into being, just as Cicero pointed out that the process of dying cannot be studied when death "pertains neither to the living nor the dead": "We are studying attentively the genesis of animals from the egg, lo! in the egg itself we behold the animal already almost formed, and our labor thus is rendered fruitless."[56] Examining fertilized eggs at intervals of six hours, Malpighi began to see the spinal cord, the primordia of the brain, and only a little later the head, neck, thorax, the beginnings of the eyes, and something like the motion of the heart.

This set of studies, evidently conducted with a good single-lens microscope, represented a considerable advance on Harvey, who saw the heart first as a pulsating point, and that only between the fourth and fifth day; it gave some support even to Fabricius's claim that the skeletal framework precedes the appearance of the fleshy organs. There is no hint of the doctrine that animals exist preformed in preexistent seeds in the parental body. Malpighi's belief was rather that females in general produce eggs and cast them off, and that these eggs, when fecundated by the male, "unfold into a new life."[57] Seeds, like eggs, he had concluded by the time of the *Anatome plantarum*, are produced by the plant or animal de novo.[58]

However, the question is not just what Malpighi believed, but what

---

[55] Malpighi, *De formatione pulli in ovo* (1673), in Adelmann, *Marcello Malpighi* 2:947.
[56] Adelmann, *Marcello Malpighi* 2:935.
[57] Malpighi, letter to Spon, in ibid., p. 861.
[58] Malpighi, *Anatome plantarum*, in Adelmann, *Marcello Malpighi* 2:849.

he said that could have stimulated the development of a doctrine at variance with his own views. His finding that more of the structure of the animal was present from a much earlier stage than had been imagined was a correction of Harvey and as such a challenge to epigenesis; it was no longer possible to maintain that the heart and the blood were the architects of the body. Malpighi describes his attempt to reveal more of what he describes as an "animal" in an unincubated egg, laid the previous day, in which he saw a fetus with a head and a carina or rudimentary skeleton enclosed in a membrane: "I have often opened the follicle with the point of a needle to release the animal confined there, but to no purpose, for it was so mucous and so very tiny that in every case it was lacerated by a light touch."[59] He concluded from his trials that "it is therefore proper to acknowledge that the first filaments of the chick pre-exist in the egg and have a deeper origin, exactly as the [embryo] in the eggs of plants."[60] Another somewhat surprising passage, as Howard Adelmann concedes, might even be taken to indicate that Malpighi believed that the chick is formed before fertilization: "in the hen's egg Nature does not scatter and sprinkle the semen of the cock . . . upon the cicatrix alone, in which the rudiments of the parts lie concealed, but moistens with plastic force the entire egg."[61]

A careful reader might nevertheless conclude that Malpighi did not believe in miniaturized presence. But no reader could fail to be influenced by the language of compendia and epitomes that appears throughout his writings. The body, he says, is "identical and similar to itself"; smaller patterns of organization are reproduced in larger ones, and this replicative formal unity is what produces the sensory unity of the whole, "because the smallest part has everything that the largest has."[62] The word "compendium," as Adelmann remarks, appears frequently in the *Anatome plantarum*. Malpighi says that the bud is "a compendium of the not-yet-unfolded plantlet"; the seed is a compendium of the entire future plant.[63]

Indeed, Malpighi sometimes preferred to make a direct seed-fetus comparison, bypassing the egg. "The seed is the fetus, in other words a true plant, with its parts . . . completely fashioned."[64] Part of the logic involved is that both seed and fetus need a surrounding matrix: the earth or the womb. He tended to see the egg as simply another such

[59] Malpighi, *De formatione pulli in ovo*, in Adelmann, *Marcello Malpighi* 2:945.
[60] Ibid.
[61] Malpighi, letter to Spon, in Adelmann, *Marcello Malpighi* 2:861–62.
[62] Malpighi, *De viscerum structura*, in Adelmann, *Marcello Malpighi* 2:843.
[63] Malpighi, *Anatomes plantarum idea*, in Adelmann, *Marcello Malpighi* 2:902.
[64] Adelmann, *Marcello Malpighi* 2:845.

repository for the fetus. But nothing depended on such identifications. The plant seed, he says elsewhere, "is an ovum which contains a fetus made up of its more important parts and which can be kept fertile even for years, until its parts are caused to unfold by the swelling of fluid entering it from the outside."[65] And this is so in the case of animals: there is no Harveian self-formation. "These and other facts seem to prove that there is present in the cicatrix a compendium of the animal, by which I mean the first outlines of the principal parts, or in other words, their outermost boundaries, which through the agency of growth become visible when motion has been communicated to the fluids and the cavities gradually fill and become turgid."[66] The egg-seed analogy was broadly convincing. "In both," says an anatomist writing for *Philosophical Transactions* in April 1683, "the Parts of the *Embryo* are designed and drawn out, before the *Eg* has been at all affected by the *Masculine-Seed*, or the *Vegetable Seed* put into the *Womb* of the *Earth*." The figure of the plant is in its seed; the figure of the chicken is in its yolk.[67]

But one could not explain the attraction of the doctrines of preexistence and preformation without noting the role played by the study of insect metamorphosis. The observations made on embryos were simply inconclusive. It became essential therefore to interpret them in light of the more definite and striking results obtained from the other realm of encased development: the transition from one life stage to another in the lower forms.

The analogy between the silkworm in its cocoon and the fetus in the womb was a compelling one, given the wormlike appearance of the tubiform early fetus in all species, and the protective covering around both. It was natural to suppose that the same process was occurring inside one as in the other—the emergence of visible form from subvisible form. Malpighi found that the pupa of the caterpillar when unraveled reveals with great clarity the outlines of wings and legs and the antennae of the imago, which itself seems to have been only rolled up and hidden, so that, just as with the egg, "its nature comes to view sooner and is more deeply rooted than is commonly believed." And indeed, it seems as though the pupa is "nothing but a mask or covering for the moth already engendered, so that if not aroused or harmed by blows from external objects, it may in this disguise acquire solidity and grow up like a fetus *in utero*."[68] He goes on to compare the compen-

---

[65] Malpighi, *Anatome plantarum*, in ibid., p. 855.

[66] Malpighi, *Opera posthuma*, in Adelmann, *Marcello Malpighi* 2:867.

[67] "An Account of the Dissection," p. 187.

[68] Malpighi, *De bombyce*, in Adelmann, *Marcello Malpighi* 2:844.

dium of the bark primordium in trees with the parts of the butterfly lying concealed in caterpillar and pupa.[69]

Jan Swammerdam, who was carrying out his researches on the silkworm at the same time as Malpighi, published his results in the *History of Insects* (1669) a few months later. He quoted Malpighi's remark that the pupa is nothing but a "mask" or "covering," and stated that he absolutely rejected Harvey's "imaginary metamorphosis," conceived as the changing of one sort of matter into another. By devising a technique for soaking and unraveling the integuments of the pupa, Swammerdam revealed the nymph form wrapped up in the caterpillar, and the butterfly in the nymph, showing that the larva, pupa, and imago are simultaneously present and nested within each other:

> The Nymph, or Chrysalis, may even be considered as the winged animal itself hid under this particular form. From whence it follows, that in reality the Caterpillar, or worm, is not changed into a Nymph or Chrysalis; nor, to go a step further, the Nymph or Chrysalis into a winged animal; but that the same worm or Caterpillar, which, on casting its skin, assumes the form of a Nymph or Chrysalis, becomes afterwards a winged animal. Nor, indeed, can it be said that there happens any other change on this occasion, than what is observed in chickens, from eggs which are not transformed into cocks or hens, but grow to be such by the expansion of parts already formed. In the same manner, the Tad-pole is not changed into a Frog, but becomes a Frog, by an unfolding and increasing of some parts.[70]

Though Harvey is referred to as an "immortal," these passages constitute a polemic against him. "The experiments we have made have," Swammerdam says, "like the rising sun, dissipated this thick and dark cloud of imaginary metamorphoses, the whole truth thereby appearing in the clearest and most evident light."[71]

Swammerdam's argument appears to be that, as generation in the hen's egg does not involve transformation, but only expansion and unfolding, the conclusion is warranted that there is no true metamorphosis in insects either. However, these passages were almost certainly read as having the reverse significance. The preexistence of the imago, which could be directly experienced, provided a model for the state of affairs

[69] The iatromechanist Giorgio Baglivi understood development, growth, and metamorphosis as a kind of sprouting that takes place under the influence of heat and fermentation: "The seeds of Teeth lie hidden in the Gums for many years, like as the Curls of Hair lie hid in their root under the Skin, until at length the necessary maturity drawing on, they break forth, as it were by vegetation." Letter to Nicolas Andry, 14 July 1699, in Andry, *Breeding of Worms*, p. 246.

[70] Swammerdam, *Book of Nature*, p. 2.

[71] Ibid., p. 9.

that could not be experienced in the hen's egg. In a curiously brief section entitled "Man himself compared with insects and with the Frog," Swammerdam extends his conclusions to human reproduction, saying that

> it is clearer than the light at noon, that man is, like insects, produced from a visible egg [and that] man, that rational animal, finds his first nourishment and represents, as it were, a Vermicle or Worm, or to use Harvey's words, a Maggot lying in the egg. It is very clearly observed, that these parts of the Man-Vermicle grow by degrees into a head, thorax, belly and limbs. . . . It is indeed very admirable to observe, how the limbs sprout about the shoulder blades, and at the lower parts of the body: for, in the beginning they resemble the small cups of flowers, or the bags and cases of the parts of insects; the former enclosing the flowers, and the latter the wings; and then, by degrees, just as the legs of the Frog, they grow out of the body and are divided into joints.[72]

The identification of fetal man with a worm was an old one and raised certain philosophical problems of speciation and identity. Thomas Moffett had felt it was necessary to insist that the worm is a man at least potentially, not a creature of another species. "[A] worm does not dye, that a [Butterfly] may be bred; but adds a greater magnitude to its former body and feet, colour, wings; so life remaining, it gets other parts and other offices: so the off-spring of man, after some daies at first of a man *in posse* is made a man actually."[73] The denial of metamorphosis in favor of a theory of the sprouting of covered-up parts in Swammerdam became finally the full-fledged doctrine of *emboîtement* in his *Miraculum naturae, sive uteri muliebris fabricata* (1672). Here he states that Adam and Eve held all future generations in their loins and repeats that "there is no place in the nature of things for the generation of parts, but only for their propagation or growth, where all accident is excluded."[74]

Daniel Le Clerc and Jean-Jacques Manget, who edited the most important collection of anatomical writing of the period, the *Bibliotheca anatomica*, suggested that this idea had actually been supplied to Swammerdam by the philosopher Malebranche, in whom we see the philosophical development of the idea and its integration with metaphysics and theology.[75] They supplied no direct evidence, however. Malebranche was at the time making his own observations in artificially incubated chicken eggs, in which, according to a colleague, "he saw the

---

[72] Ibid., p. 104.

[73] Moffett, *Theater of Insects*, p. 1040.

[74] Swammerdam, *Miraculum natuare*, in Adelmann, *Marcello Malpighi* 2:908.

[75] On Malebranche and Swammerdam, see Le Clerc and Manget, *Bibliotheca anatomica* 1:497; see also Fouke, "Mechanical and 'Organical' Models," and Adelmann, *Marcello Malpighi* 2:907.

heart formed and beating and a few arteries."[76] This Harveian discovery was familiar to every preformationist and epigeneticist alike and has no particular bearing on the doctrine of *emboîtement*.

The inference to epigenesis was in any case overridden for Malebranche again by a change of field. In *The Search after Truth* (1675), he cites the equally familiar observation that the seed of a tulip bulb examined in the dead of winter can be shown to contain the rudiments of leaves and flowers, and he promptly generalizes to apply this to all plants and animals. "Nor does it seem unreasonable to believe even that there is an infinite number of trees in a single seed, since it contains not only the tree of which it is the seed, but also a great number of other seeds that might contain other trees and other seeds, which will perhaps have on an incomprehensibly small scale other trees and other seeds and so to infinity."[77]

We have here an exact formulation of the theory that complete organisms preexist in all living creatures. But how was this confidence sustainable when in Malpighi's published drawings, as well as in Malebranche's own ocular experience, the ten-day-old embryo looks like nothing more than a kind of tube with huge eyes at one end? Malebranche says, less than categorically, that "a chicken that is perhaps entirely formed is seen in the seed of a fresh egg that has not been hatched, this germ being under the tiny white spot—the cicatrix on the yolk," adding that frogs can be seen in frogs' eggs and that "still other animals will be seen in their seed when we have sufficient skill and experience to discover them."[78] The discrepancy nevertheless troubled him. In the later *Dialogues on Metaphysics*, which devotes an entire chapter to preformation, Theotimus admits that we cannot see every part of the plant in the seed, but "we may try to imagine them." Indeed, we are obligated to try to imagine them: reason demands that we envision what we cannot see. No one will think simplemindedly that the bee, for example, contained in the larva from which it will emerge, "has amongst its organic parts the same proportions of size, solidity and configuration as when it emerges." Everyone admits that "the head of a chicken when it is in the egg and appears as in the form of a worm is much larger than all the rest of the body and the bones assume their proper consistency only after the other parts. I claim," he says, "only that all the organic parts of bees are formed in their worms and so well-proportioned to the laws of motion that, through their own construction and the efficacy of the laws, they can grow and take on the shape fitting their estate without God lending a hand, as it were, through

---

[76] Letter from D. David to N. Poisson, 10 August 1670, in Malebranche, *Oeuvres complètes* 19:789.

[77] Malebranche, *Search after Truth*, p. 27.

[78] Ibid.

extraordinary Providence."[79] The true chicken appears in the form of a worm!

With this concession made, we may wonder whether the boundary lines between preformation and its alleged rivals were not falling back into the indistinctness from which they emerged in the highly synthetic treatment by Gassendi. What is the difference between saying that the whole organism preexists, but not in the form in which it will later appear, and saying that it exists only potentially, or that some precursors of its parts are present? Malebranche's overriding concern here as elsewhere is to account for the phenomena by reference to laws of nature that are simple, productive, and universal—that is to say, mechanical—without reference to forms, forces, or inner agencies. The demand for a preformed organism is thus relative to the demand that the ordinary laws of nature be able to produce a new being through their regular operation alone. Malebranche regards this as impossible unless the task is to some extent achieved in advance. No reasonable person can accept the Cartesian hypothesis that mechanical laws alone are sufficient to produce organized form: such laws, he says, may explain the production of the entire cosmos as we see it from the chaos of the primeval vortices, but it is too much to ask that, from the chaos of any reproductive primordium, a complete animal should emerge. Thus the theory of preformation could be better defined by reference to what it excluded—scholastic potentials, incorporeal agents—than by reference to what precisely it committed the holder to, or what selection of optical experiences it would have taken as corroborating or refuting it. To this extent it is correct to suppose that it represented the "rationalist" position. This judgment should not, however, force the historian into the awkward position of taking Harvey's spiritistic epigenesis as empiricism. In general it can be said that those farthest from the sphere of direct observation—and here one would have to include Descartes, Malebranche, and Leibniz—were the most coherent and definite in their assertions, while those with a true intimacy with their subjects allowed the theory to float. Malpighi was thoroughly opportunistic and philosophically inconsistent in his explanatory apparatus, employing now the terms of Cartesian mechanism, which would explain growth as a process of fluid and particle accretion, now the language of plastic powers and unfoldings, as each seemed suitable.

The visual imagination then played a smaller role than the force of analogy and the plausibility of extrapolations. This is not to say that microscope-induced illusions and outright hallucinations in early embryology were unknown. Pierre Borel saw with his microscope "a whole little chicken" in an egg a few days old, as well as the rudiments of plants in seeds in 1655. William Croone of the Royal Society achieved notoriety in 1679 for observations performed earlier on chicken eggs, which he thought refuted Harvey. His account is full of the circumstantiality prized and expected by the society. A visit to a farm, a conversation with the farm wife, and an encounter with a hen provided Croone with a freshly laid egg that he took home and left near his hearth, returning to find it semicooked. Upon opening it, he saw a small sac adhering to the yolk which he separated and placed into water. "Next I scanned with eager eyes all there was to see, and observed quite clearly and distinctly those two rather large vesicles which are the eyes, with the beak lying in between. I saw besides little ribs of a milky colour, and the rudiments of the feet were just showing."[80] A second experiment on a forty-eight-hour-old embryo revealed even more in the way of internal organs. "I tore away the second membrane, whereupon the tiny body [the cicatricula] with all its equipment of vessels came clearly into view." Underneath the cicatricula was hidden the body of the chick, with head, wings, tail, liver, spleen, and feet.[81] Croone was sure that the existence of the heart implied the existence of a brain, for the former "cannot produce regulated pulsation without the help of the brain and nerves," so that a nervous system must exist from the beginning "even if this belief is not confirmed by observation."[82]

Croone's report to the society met few of their own tests of reliability. First, he was reminiscing on experiments carried out by him approximately seven years earlier; second, he was not sure that all eggs were like this—"in no case," he admits, "did I observe in the colliquament the beginning of the chick taking place always in a uniform manner"; finally, he was not sure that any of his observations would prove replicable: some people might, he allows, "consider the trustworthiness of these observations as extremely slight or else conclude that I had been dreaming."[83] Nor were these highly tentative conclusions based on microscopical observations; he had initially tried, he says, to look at freshly laid eggs with a lens, but "I perceived that I was gaining no advantage but that by its aid nothing was presented to my eyes which I did not see without it." Croone's excesses, then, while they might have contributed to the impression that embryology was a confusing and difficult subject, proved nothing about the unreliability of early microscopes or their contribution to overinterpretations.

The observations and interpretations of Nicholas Steno, who looked at the "testicles" of viviparous female fish and determined that they were

---

[80] William Croone, reproduced in Cole, "Dr. William Croone on Generation," p. 121.
[81] Ibid., p. 125.
[82] Ibid., p. 127.
[83] Ibid., p. 134.

ovaries, which he reported in his *Elementorum myologiae* of 1667, were confirmed within the space of five years by the Dutch anatomists Jan van Horne, Theodore Kerckring, and Regnier de Graaf, the last-named demonstrating "eggs" in rabbits, cows, dogs, and swine. The theory was advanced that the ovary of every female produced eggs that fell somehow into the uterus, were, at some point, fecundated, and grew into embryos. Harvey's fundamental insight was thus confirmed, despite his rejection of a visible mammalian egg, and *Ex ovo omnia* took on a more definite meaning. There were some difficulties with this account. The "eggs" that de Graaf showed in table 16 of his *De mulierum organis generationi inservientibus* of 1672 were too large to pass through the fallopian tubes into the uterus, and he found that the rabbit embryo on the third day after copulation was still smaller than the object identified. He solved this paradox by insisting, correctly, that the true ovum was not that object but rather an entity ten times smaller, surrounded by nutritive or other matter.[84] The fecundating role of the semen on the egg was now seen characteristically in magnetic or chemical terms. Borelli had argued that the semen exercised an orienting function, as iron, when rubbed with a magnet, is "given direction and acquires magnetism, or the magnetic power to determine the poles of the earth."[85] De Graaf "constantly asserted," according to Leeuwenhoek, that the semen is the "Vehicle of a certain Volatile Salt, or such like spirit, conveying to the Egg of the Female a *Contactum Vitalem*."[86] The alchemical idea that salt is the principle of form and fixity perhaps lay behind this suggestion. A third alternative was fermentation, favored by Descartes and his disciple Pierre Sylvain Regis, who believed that the development of the preformed ovum was set in motion by a chemical reaction that causes ebullience and expansion in the mass.[87]

A few extraordinary claims were made by established and respectable anatomists. Kerckring, who specialized in the study of aborted embryos and who was once lucky enough to get hold of a female subject killed in flagrante delicto, announced in his *Observationes medicae* (1672) that he had found "eggs like birds'" in all the females he had studied, including virgins, and said that he had seen in a human embryo that he believed to be only three or four days old already the head and traces of organs, although the eyes, nose, and mouth did not become visible until the fourteenth day. But such claims were still well within the range of

---

[84] De Graaf, *De mulierum organis*, p. 193; cf. p. 315.

[85] Adelmann, *Marcello Malpighi* 2:841.

[86] Leeuwenhoek, "Concerning Generation by an Insect," p. 1121. In *De mulierum organis*, de Graaf describes semen (as Daniel Sennert had) as a "spiritous, warm, moist body arising in the testicles suitable for the propagation of the soul" (p. 124).

[87] Regis, *Cours entier de philosophie*, book 8, part 3.

what had previously been reported and consistent with prevailing assumptions. The egg theory was, in these years, "almost universally received" according to a *Philosophical Transactions* survey in 1683, despite lingering worries about the size of the egg and its passage from the ovaries to the uterus, and Harvey's reputation was upheld. The two-fluids theories of Descartes and the Galenists were regarded by most as thoroughly discredited.

The discovery of the spermatozoa later in the 1670s was for this reason unwelcome. Forty years later, after Leeuwenhoek's death, the vice president of the Royal Society, Martin Folkes, described it as having given "a perfectly new Turn to the Theory of Generation, in almost all the Authors that have since wrote upon that Subject."[88] But spermaticism remained a minority position well into the first quarter of the eighteenth century. Leeuwenhoek's vigorous representation produced a flurry of interest but little conviction except in the popularizer Nicolas Andry de Boisregard; Nicolaas Hartsoeker turned away from it, and even Leibniz, who was friendly to the idea, was never entirely convinced. Though Leibniz later joked that Leeuwenhoek had thereby restored the preeminence of the male sex, this was small consolation for the confusion into which the theory of generation, having made such progress with respect to the egg, was now thrown. True, some women had been insulted by the comparison of themselves with fowls, and the implication that women in holy orders and young virgins had an indelible sexual destination imprinted on them despite all wishes to the contrary left some observers of the intellectual scene uneasy. In fact the ideological value of ovism was unstable; theologians could not agree whether women were degraded or excessively elevated by it.[89]

The discovery of the spermatic animalcula was also attended at first by shame, presenting the transformation of private into public knowledge and the scientific exposure of the intimate in an interesting light. According to Leeuwenhoek's own account, Oldenburg had requested in a letter to him of 24 April 1674 that he examine "saliva, chyle, sweat, &c.," and Leeuwenhoek had taken this "&c." to mean that he should examine semen too; he found, however, only "globules." "[A]s I felt averse from making further inquiries, and still more so from writing about them," he explained later, "I did nothing more at that time."[90] Leeuwenhoek was not to any degree squeamish; he was rather morally concerned lest anyone should think he had produced his samples by

---

[88] Folkes, "Mr. Leeuwenhoek's Curious Microscopes," p. 449.
[89] See Roger, *Les sciences de la vie*, pp. 263, 267.
[90] "Observationes D. Anthonii Lewenhoeck," p. 1042.

"sinful contrivance," rather than, as he testified, by hastening from the marriage bed to the microscope. In 1677, however, a better opening for communication occurred when a young researcher named Johan Ham van Arnhem visited him with a small glass phial containing "the spontaneously discharged semen of a man who had lain with an unclean woman and was suffering from gonorrhea; saying that, after a very few minutes . . . he had seen living animalcules in it which he believed to have arisen by some sort of putrefaction. He judged these animalcules to possess tails, and not to remain alive above twenty-four hours. He also reported that he had noticed that the animalcules were dead after the patient had taken turpentine."[91]

At this stage of the letter, which was sent in 1678 and printed the following year, Leeuwenhoek felt safe in mentioning that he had been making similar observations on semen, and these in healthy rather than sick men, and that he had seen

> so great a number of living creatures in it, that sometimes more than a thousand were moving about in an amount of material the size of a grain of sand. . . . These animalcules were smaller than the corpuscles which impart a red color to the blood; so that I judge that a million of them would not equal in size a large grain of sand. Their bodies were rounded but blunt in front and running to a point behind, and furnished with a long tail, about five or six times as long as the body, and very transparent, and with the thickness of them about one twenty-fifth that of the body. . . . The animalcules moved forward with a snake like motion of the tail, as eels do when swimming in water: and in the somewhat thicker matter, they lashed their tails some eight or ten times in advancing a hair's breadth. I have sometimes fancied that I could even discern different parts in the bodies of these animalcules; but forasmuch as I have not always been able to do so, I will say no more.[92]

Apparently Leeuwenhoek had observed these creatures carefully from the earliest period of his researches and had wondered whether they were not early forms of the human being. However, in his letter to the society, he suggests that his attention was not focused on the animalcula but on the fluid in which they swam. Surprisingly, he reports that the "denser substance" of the semen was seen to be made up of "all manner of great and small vessels, so various and so numerous that I misdoubt me not that they be nerves, arteries, and veins. . . . And when I saw them, I felt convinced that, in no full-grown human body, are there any vessels which may not be found likewise in sound semen."

This double communication prompted a request for further study,

[91] Ibid.
[92] Ibid., p. 1041.

but also a protest of irrelevancy. The new secretary of the Royal Society, Nehemiah Grew, wrote back to ask him to extend his observation to animals other than man and expressed serious doubts that the semen itself could be differentiated as Leeuwenhoek said. Grew was troubled by the threat to what seemed like established facts about the female egg and its role and suggested that Leeuwenhoek had been under some kind of illusion. Harvey, he pointed out, had never found semen in the uterus after coitus; and de Graaf had shown that eggs pass from the fallopian tubes into the uterus and there develop into the fetus. The reasonable conclusion to draw was that, as he says, repeating de Graaf, "the male semen is nothing except the vehicle of a certain breath, most volatile and active, impressing upon the conception, i.e., upon the female ovum, its vital touch."[93]

Leeuwenhoek replied on 18 March 1678, sending drawings of the spermatozoa of a dog and a rabbit. Meanwhile, confirmation was appearing from other quarters. Huygens reproduced his results after reading a letter of Leeuwenhoek's in December 1677 and announced them to the Paris Académie des Sciences in July 1678. Nicolaas Hartsoeker began to write in that year of his studies of the semen of other animals; he would claim in his *Essai de dioptrique* (1694) that he had actually been the first to see the spermatic animalcula but had thought it indelicate to announce this discovery to the world (he said that he told people that he had found animalcula in his saliva), and there is perhaps less reason to doubt his word here than those eager to give priority to Leeuwenhoek will concede. Leeuwenhoek's discovery was disseminated by Pierre Bayle in the *Nouvelles de la république des lettres* of 1684, where he published a Latin (so as not to offend ladies) translation of Leeuwenhoek's letter to Christopher Wren of January 1683, and Jean Le Clerc discussed animalculism in the *Bibliothèque universelle* of April 1686.[94] But Bayle was still leaning, with the majority of learned France, toward ovism in 1685.

Ham's first thought had been that the spermatic animalcula were a consequence of his subject's venereal infection. But this interpretation could be quickly ruled out, as all healthy men and male creatures in general seemed to have them. By 1699, Nicolas Andry had shown that they were not present in the prepubescent, the old, or the very ill, thereby strengthening the probability that they were in some way involved in generation. But what was this role? If they were indeed actual little humans, where did they come from, and how could the deaths of most of their vast numbers be justified in providential terms? Leeuwen-

[93] Ibid., p. 1043.
[94] Roger, *Les sciences de la vie*, p. 311.

hoek believed in 1683 that the animalcula were divided into male and female (he thought he had seen some difference where the tail joined the body)[95] and perhaps reproduced by copulating, and he maintained this view—less strange than it sounds in view of his discovery that parasitic worms in animal bodies had sexual organs—for years.[96]

Because males produce such a quantity of sperm, spermaticist *emboîtement* was a less likely hypothesis than the continued production of animalcula. But recourse to spontaneous generation was a catastrophic conclusion: if the animalcula were animals, preexistence seemed to imply its own negation! Leeuwenhoek's early notion that the animalcula came into being through their own sexual congress only pushed the problem back into further regions of smallness. But there was eventually no doubt for him that the animalcula were produced and did not preexist, for, while female eggs—which Leeuwenhoek thought of as either a source of nourishment and shelter or emunctories for certain tiny vessels[97]—indeed seemed to be present in females at birth, more animalcula are produced by a male over the course of a lifetime than (he calculated) could be kept inside the body, fish especially producing vast amounts of spawn every year. Further difficulties were raised for him by the case of aphids, the only animals he could find in which there seemed to be females only and reproduction without copulation by parthenogenesis.[98] This was impossible on his theory that the female is only a nidus, and his interim solution was apparently to treat the aphid itself as the logical equivalent of a spermatozoon, and to suppose that it reproduced asexually: the resulting picture was one of spermaticist *emboîtement.*[99]

Leeuwenhoek maintained vigorously the identity of the animalcule and the later fetus. "It's supposed," he wrote in his letter of 30 March 1685, "that Animals while they grow in the Womb, have a living Soul; now if this be so, it's a thousand times more likely, that the Soul which is in the *Animalia Seminis Masculi* should still remain there, changing only their outward shape, then [*sic*] that this should pass into another body."[100] He did not, however, set much store in observed differences between different species of animalcula. He complained that the view that the sperm of dogs and cocks contains little puppies and pullets had

[95] Leeuwenhoek, "Letter to Sir CW," p. 75; see also Leeuwenhoek, "Concerning Generation by an Insect," p. 1127.

[96] See Cole, "Leeuwenhoek's Zoological Researches," p. 10.

[97] Leeuwenhoek, "Letter to Sir CW," p. 75.

[98] Cole, "Leeuwenhoek's Zoological Researches," p. 13.

[99] Ibid. By 1713, he had reverted to the view that aphids are females.

[100] Leeuwenhoek, "Concerning Generation by an Insect," p. 1126.

been mistakenly ascribed to him.[101] His comparative illustrations published in *Philosophical Transactions* in 1685 showed instead the surprising similarity in the shapes of dog, rabbit, and human spermatozoa. He distanced himself soberly from other preformationist claims. The microscope does not reveal a tree in the kernel of an apple, but only "roundish transparent globules"; however, we do see at some point leaves, ribs, and vessels in the sprouting seed. But in a letter of 1682, he had stated that "The *Humane Foetus*, tho' no bigger than a *Green Pea*, yet is furnished with all its parts,"[102] and in 1685 he said that the spermatic animalcula were complete animals: "If an Insect has limbs and Bowels, tho it is not a thousand Millionth part so big as a sand [grain], then may not the *Animalia Seminis*, have likewise the same Limbs and bowells, which the *Foetus* has when it is brought forth?"[103] And he did not hesitate to broadcast some of his own sightings of early fetal, as opposed to spermatic, life. He found, after several trials at preparation involving soaking specimens in oil of turpentine, that, in what had at first appeared to be only a reddish fleshy substance adhering to the side of an ewe's uterus, he could find rudiments of the brain, eyes, and muscles, and blood vessels in a space one-eighth the size of a pea, seventeen days after coitus.[104] He also claimed to have seen in the uterus of another ewe three days after coitus a little particle the size of a grain of sand. According to Henry Baker, after examining it "with an excellent *Microscope*, he with great Pleasure found it to be an exceeding minute Lamb, lying round in its Integuments, and could plainly discern its Mouth and Eyes."[105]

Grew's appeal to Harvey's claim that no trace of male semen had ever been found in the womb was not successful. Leeuwenhoek was able in 1685 to cite an experiment in which a mated female dog was killed by running an awl into her spinal medulla. Nothing was seen in the uterus by the naked eye, but microscopical examination revealed spermatic animalcula in its "horns," that is, farther up in the fallopian tubes, indicating that they had passed through it.[106] Still, the invariable presence and huge number of spermatic animals seemed to many to argue against their having a role in generation. Though it had been appreciated that the number of eggs in the females of a species was also

---

[101] This accusation was allegedly raised in the Amsterdam *Collectanea Medico-Physica*, but I have not been able to find it.

[102] Leeuwenhoek, "Generation by an Animalcule," p. 349.

[103] Leeuwenhoek, "Concerning Generation by an Insect," p. 1128.

[104] Ibid., p. 1127.

[105] Ibid., p. 1121.

[106] Ibid., pp. 1121–22.

very large—Henry Baker would eventually count 12,444 in one crab—
this was a small number compared to the millions of spermatic animalcula in one drop of semen. Leeuwenhoek did not see this as a difficulty: just as trees overproduced blossoms and seeds and small seedlings
competed for existence, so there was competition among animalcula.[107]
Cruel as this doctrine might seem, one could avoid its harsher implications by imagining, as Leibniz did, that the spermatic animalcula were
ensouled but elevated to full human status only when they began to
grow into fetuses, when God, according to the account presented for
approval to the theologian Bartholomew Des Bosses, miraculously adds
a degree of perfection entailing rationality and reflectivity to their previously appetitive and sensitive souls.[108]

Andry, who discussed the spermatozoa in his general treatise on
worms, defended animalculism in part by arguing that their shape was
decisive evidence for their role. "The Spermatic Worms in a Man," he
notes, "have a head much bigger than the Spermatic Worms in other
Creatures, which agrees with the Figure of the *Foetus* or human Birth,
which when it is little, seems to be no more than a Great Head upon a
long Body, that seems to end in a kind of a Tail."[109] Like insects, these
worms may shed their coats as they develop:

> Does not the thing seem to speak of itself, and to tell us plainly that Man, and
> all other Animals, come of a Worm; That the Worm is the Epitome of the
> Animal that is to come of it; that if the Worm be Male, it produces a Male;
> and if it be Female it produces a Female; that when it is in the Matrix, there it
> takes its growth by means of an Egg into which it enters, and where it stays
> the time appointed by Nature; when it is grown to a certain measure, it forces
> the Membrane of the Egg and is born?[110]

This was plausible reasoning. The inertness and undifferentiated form
of the fertilized egg and its contents in its early stages, by contrast with
the vigorous motion and strikingly animal form of the spermatozoa,
might leave us surprised that, despite the major difficulties that faced
it—the origin of the spermatozoa, their moral-theological status, and
the apparent contribution of females to heredity—spermaticism did not
win greater acceptance. But the discovery of microparasites and the
widely observed animation of fluids, often with flagellate microorganisms, tended to offset the likelihood of the claim that these animalcula
had a special role in the creation of human beings, that they were human beings. These same facts, parasitism and pan-animism, clouded the

[107] Leeuwenhoek, "Letter to Sir CW," p. 76.
[108] Leibniz, *Letter to Des Bosses*, 30 April 1709, in *Philosophische Schriften* 2:371.
[109] Andry, *Breeding of Worms*, p. 178.
[110] Ibid., p. 183.

question of animate contagion as well. Ham had first interpreted the spermatozoa as a manifestation of illness, and Andry, good spermaticist though he was, was still, almost twenty years later, discussing them as though they were disgusting pests: "If you cut up a dog, and after you have taken off one Testicle, by the help of a good microscope you examine the Humour that comes out of the different vessels, you shall discover in it such a hideous number of little worms that you shall hardly be able to believe your own eyes."[111]

So much, then, for the new turn in the theory of generation. J. F. Ortlob in his *Historia partium et oeconomiae humanis* of 1692 refers to the "singular theory of generation of Leeuwenhoek." The respected Vallisnieri took up the cause of ovism in his letters and in his *Istoria della generatione*, published in 1721, and mounted a lengthy attack on the competence of Nicolas Andry. French natural philosophers especially denied that there were worms in living semen, as opposed to corrupted and deteriorating semen, at all, or denied that they were other than ordinary parasites. As for Leeuwenhoek himself, it was proof of his stubborn independence—the same independence that made it possible to discount his animalcula—that, as he told Leibniz in 1716, he did not care how many were against him; he knew he was in the right. Leibniz's famous stated preference for a Leeuwenhoek who would tell him what he saw over a Descartes who would tell him what he thought left him in a predicament when faced with a Leeuwenhoek telling him what he thought.

To carry the subject further into the eighteenth century is to follow the solidification of the doctrine of preformation in Charles Bonnet, following Malebranche and Leibniz. Bonnet argued again that the very notion of the organism as a machine required the simultaneous existence of all its parts: the heart cannot beat without the stimulation of the brain or pump blood without there being a circulatory system to pump it through. If organs are there, a nutritive system must be there to support them, and if movement is possible, so is sense.[112] Albrecht Haller wavered between epigenesis and preformation, deciding for preformation as late as 1758. Haller could not, in his preformationist phase, imagine how epigenesis could produce differentiated individuals but constant species, why peacocks should produce peacocks and chickens chickens. He saw that adding spirits of wine to what looked at first like a homogeneous embryonic jelly revealed fibers, vessels, viscera, and so on, and decided that the parts were present but transparent and soft, and thus

---

[111] Ibid., p. 179.
[112] Bonnet, *La palingénésie philosophique*, p. 97.

scarcely visible.[113] Caspar Friedrich Wolff's observations restored the credibility of epigenesis, reviving a not unwelcome vital force.[114] Phenomena that had been difficult to explain on the preformationist view, especially the regeneration of bodily parts in the more plastic species, such as the crayfish studied by Hartsoeker and the polyp studied by Trembley and Baker that "reproduced" by budding, seemed to argue for such a force. And these studies on particular creatures were suggestive in a way that the claims for the existence of a universally acting formative faculty—the much-maligned antiquarian "plastic nature" of the Cambridge Platonists—had not been. But Wolff did not refute preformation; even in the late nineteenth century, epigenesis and preformation furnished two poles of reference, and the degree to which the form of the future organism is specified in advance and the degree to which it emerges through interaction with the local environment remains a subject for experimental study.

The time elapsing between Harvey's exhaustive study of the egg, in which macroscopic observation is supplemented by an interpretation of the emergence of form stated in terms of immaterial *invisibilia*, and the identification of the mammalian ovum (or at least the place where it was to be found), the male spermatozoon, and the features of the very young embryo, was short: little more than twenty-five years. The standard treatment of preformation as a speculative theory exemplifying the domination of expectation over perception, as an example of the anticipation of nature that Bacon warned against, obscures the importance of the visualization of what had not been anticipated. The seeing of the unexpected was the primary result; conceptualization in terms of the plausible and the inherited was its sequel.

Perhaps it is true that Harvey not only could not find but did not want to find anything in the uterus of the doe or any other vivipara, that he did not want to observe the details of generation, that the "labyrinth of efficient causes" was one he did not want to negotiate. His own account, in which imagination, love, desire, warmth, and attraction spin a kind of semantic net around the subject, was one that the discourse of materiality could not invade.[115] In his watching and waiting, in his destruction of categories of difference handed down from antiquity (viviparous and oviparous, active and passive), in his ability to put down a record of what he saw, Harvey presents us with one face of the objectivity of science, one for which he was recognized and honored. The

---

[113] Haller, *Sur la formation du coeur*. On Haller's indecision, see Roe, *Matter, Life, and Generation*, esp. chap. 2.

[114] Wolff, *Theoria generationis*; see Roe, *Matter, Life, and Generation*, pp. 114ff.

[115] Compare Gillispie's general assessment of Harvey, *Edge of Objectivity*, pp. 72–73, with the work of Pagel, *New Light on William Harvey*.

overconfident and overgeneral mechanism of Descartes was, by contrast, nothing more than a dream of his own devising; and Gassendi's broad learning and divided allegiance to epicurean atomism and to formative faculty theories prevented him from asserting anything negative and so from asserting anything determinate. Without the microscope, however, Harvey could not be objective in the sense we understand; his analogies and correspondences, his songs, ideas, odors, and ferments, all the manifestations of immaterial agency, had to fill in for what he could not see. The preformationist interpretations of eggs and animalcula in the late seventeenth century may have been, as Roger suggests they are, an evasion of the problem of generation, an assertion of being as opposed to becoming. Here as well, where distinct vision left off, appeals to metaphysics—this time to the exhaustiveness of the mechanical order of nature and the singularity of the event of creation—began. But we need not read the theory of preformation as an example of the distorting influence of ideas on visual experience; we may see it instead as an uneasy compromise between experience and intelligibility. What should interest us is not simply the imaginary status of the tiny plants and animals "seen" in seeds and eggs by microscopists, but the acceptance and intellectual management of the microworld's unlikeness.

5

Animalcula and the Theory of Animate Contagion

REPUTABLE scholarship has it that the germ theory of disease originated with Louis Pasteur, or, more precisely, in the Italian medicine of the mid–eighteenth century that preceded his researches by a few years. The received view is that Pasteur's work on fermentation and infection from the 1860s to the 1880s marked a new beginning for the theory of animate contagion, and that it is strictly incorrect to see the germ theory of disease of Pasteur and, later, Robert Koch as the culmination of earlier suspicions about the role of microscopical living beings in producing illness.

Certainly this view is correct insofar as theory, practice, and communication were effectively combined for the first time in Pasteur's exemplary experimental manipulations and his published accounts of them. But if Pasteur's experimental protocols were typical of the standards achieved outside the biological sciences of his time and untypical only for medical experimentation, it is nevertheless difficult to agree that he is the author of a discontinuous episode in the history of science. Why the historical evolution of the theory of animate contagion should have been overlooked or denied is something of a historiographical mystery that I shall try to solve in the course of this chapter by examining the quality of evidence in the first period of microscopical investigation for the theory that disease is caused by minute living beings. It will become clear in the process that a distinction must be made between the coherent and well-motivated promulgation of such a theory and its reception by doctors and philosophers. If the acceptance of preformation and preexistence involved the philosophical nurturing of a theory strongly underdetermined by the visual experiences of microscopists, the rejection of animate contagion teaches the complementary lesson: the value of observations that generate no enthusiasm among metaphysicians is likely to be underestimated.

The theory of animate contagion survived in the eighteenth-century medical literature without being particularly cherished, although it attracted the curiosity of Carl Linné and René-Antoine Reaumur and even had a few distinguished proponents, such as Antonio Vallisnieri. Pasteur was himself extending the work of Agostino Bassi, who, in the 1830s, saw that the silkworm disease known as *muscardine*, which was threatening the textile industry, was caused by a fungus, and who extended the theory of living contagion to chicken pox, plague, syphilis, gangrene, cholera, and pellagra.[1] By marrying the theory of infectious disease to the chemistry of fermentation, Pasteur clarified the puzzling logic of good and bad infestations. The missing links in this story concern the fortunes of animate contagion between Vallisnieri and Bassi. The assertion of the Singers that "from the year 1725 nothing of real value appeared on the subject until the independent development of the theory by the group of nineteenth-century writers, of whom Pasteur is the crowning glory,"[2] is a concession that animalculist theories of disease did assume significance before 1725, so that it is their failure to achieve widespread acceptance for over a century, and the relevance of this failure for a theory of scientific progress, that needs explanation. A reading of the plague and pestilence literature of the first period of microscopical investigation is a helpful exercise.

Outbreaks of bubonic plague were severe in Europe in the first half of the seventeenth century and gave rise to a body of interpretive literature of medical and philosophical interest. Though the special character of plagues was of interest to the ancients and prompted speculation about the mode of transmission, epidemic illnesses fell outside the realm of understanding and therapeutic practice supplied by Hippocratic and Galenic medicine. Traditional medical theory took illness to be an affection of an individual body, traceable to an imbalance of humors, or a reaction to some environmental stress or overindulgence. It was not well equipped to comprehend an illness that was not only strikingly nonselective, affecting young and old, male and female, workers and the idle rich alike, but was sudden, violent, and uniform in its manifestations.

In a Christian culture, the religious meaning of plague had to be deciphered, and the plague treatises of the first half of the seventeenth century reveal the disease as a medical, theological, and political problem whose seriousness tested the limits of the society to cope in practical terms and the limits of its science to explain what the society was experiencing. In such periods of stress, animism and personification are natural conceptual resources. George Wither's poem *The History of the Pestilence* of 1625 suggests that the cause of the plague can only be what is in essence an image of the plague: the plague maiden. No mere natural and so heteronomous cause—dirt, winds, warmth, or cold—could produce a spectacle of this dimension. As the disease replicates

---

[1] Bassi, *Del mal del segno calcinaccio o moscardino.*
[2] Singer and Singer, "Doctrine of *Contagium Vivum*," p. 206.

itself and spreads through the population, so its source must be the original model of the illness itself, with her spots, her foul breath, her poisons, and her frantic look: "She tooke up all her boxes of *Infections* / Her *Carbuncles*, her *Spotts*, her *Soares*, her *Blaynes*."[3]

Wither's intention in his poem is to show how plague is both the picture and the consequence of sin. He hopes to enjoin repentance, to defeat naturalism, and to stifle human interference. The plague is punishment from God and must be accepted as such. As Thomas Dekker in *Newes from Graves-ende* (1604) explained it, "every man within him feedes / A worme which this contagion breedes; / Our heavenly parts are plaguy sick, / And there such leprous spots do stick."[4] The worm functions here as a symbol of corruption; it is bred by the disease and is not its cause. But what of real worms, or insects, or microscopical animals, and illness?

The early plague treatises hold in moral contempt physicians, who, according to the conventional understanding of their profession, are all atheists and mostly charlatans who peddle ineffective cures. The unanimity of their authors and the weight of the arguments they assemble to show that plague is of supernatural rather than natural origin is striking. Yet, by the end of the century, even though the cycle involved in its transmission from bacterium-infected flea to rat to flea to human was not clarified until the late nineteenth century, the dominant understanding of plague had shifted to a naturalistic footing. The most respected theory of the period stated that it was a form of chemical poisoning, due ultimately to the deleterious action of toxic corpuscular effluvia emitted from the earth. This theory triumphed over its rivals, which included, in addition to the theory that plague was a direct stroke of God, that it was caused by evil planetary conjunctions, that it involved a poisonous emission from the stars, that it was transmitted by an "idea" or alien *archeus*, and, finally, that it was transmitted by microscopical insects whose eggs were carried through the air, lodged in people's bodies, and there developed into animalcula, which produce horrible and fatal symptoms. The late seventeenth century's preferred conception of plague as an effluvial chemical poisoning was a conceptual achievement that brought the origin of plague in a literal sense back to earth. Plague did not come *de supra*, from somewhere beyond the world—heaven or the stars—but *ab infra*, from repositories below ground. It was not a missive from God, or a missile for that matter, but the effect of the normal operation of subvisible agents: its cause was not occult in the profound sense, but only contingently invisible to the un-

---

[3] Dekker, *Plague Pamphlets*, p. 50.
[4] Ibid., p. 85. On this literature, see esp. Wilson, *Plague in Shakespeare's London*.

aided eye. The disease itself was a process of subvisible microdestruction, working from the interior recesses of the body out to its surface, where the effects could be seen and felt on the patient's body.

Mediating between the poets and theologians on one hand and the doctors on the other were the natural philosophers who sought to allocate the responsibility for plague equally to metaphysical and physical agents. Isbrand Diemerbroeck's often-cited, reprinted, and translated treatise of 1665 claims that the cause of the plague is a venom generated directly by God, which then propagates itself "like a ferment in Leven, through the body and masse of the Aire to the destruction of Animalls."[5] He points out that God is said in the Scriptures to be its author, that its effects exceed the power of all natural causes, and that, because God created all things good at the beginning of the world, plague cannot be part of the original creation and must be a postlapsarian innovation. Plague as a manifestation of the wrath of God had its biblical warrant. The Scriptures recount how God sent sores and sickness, fire, and destructive insects as a warning and a remembrance to his people in ancient times. And the typical outbreak of plague in an urban seat of corruption where luxury, cozening, and fraud were centered suggested that the pride and greed of the cities were the immediate provocation; the rustic virtues seemed to help resist, to some degree, its spread.

That the plague was warning and chastisement the early-seventeenth-century writers did not doubt. Like Diemerbroeck with his ferment, they allowed that natural causation as well as divine was involved, but they regarded it as impiety to refer it solely to natural causes.[6] The "dull Naturallists" needed to enlarge their views; and the main argument that George Wither could bring against them was that plague killed and spared unpredictably. Its source lay "not in the Constellations, or the Weather, / for then we should be poysond alltogether."[7] The plague was rather "a Rationall Disease, wch cann / Pick with discretion here and there a man,"[8] sparing this one for his goodness or preserving him for an even worse punishment. Its higher rationality was hidden: plague could not be comprehended by the intellect. "Such knotts and Riddles; that it much amazes / The naturall man: Because he seldome findes / (As

[5] Boghurst, Loimographia, p. 14.

[6] Wither, History of the Pestilence, p. 64. As he puts it, "Shee partly Metaphisicall apeeres. / And partly Naturall; she oft may carry / Her Progress on, by means that's ordynary." But he criticizes those who put their faith in the efficacy of preventive measures or who find natural causes in "Humours, Inclynations, Coniunctions, Planetary-Constellations."

[7] Ibid., p. 53.

[8] Ibid., p. 65.

he perceives in greifes of other kindes) / The *Causes* and *Effects* agree together."[9] Whom the plague would strike, the exact constellation of symptoms, whether the victim would succumb within hours or die a lingering and painful death, were all impossible to determine. Remedies were impotent; gums, herbs, plants, roots, chemical medicine, charms, unicorn's horn were all useless, "except adioyne we do / A medcyne *Metaphisicall* thereto." That medicine was repentance, sobreity, charity, faith, and fortitude. Dekker too could reject the proposal that sickness proceeds from the air, from heaps of corpses of soldiers or accident victims, from "standing Pooles, or from the wombes / Of Vaults, of Muckhills, Graves, & Tombes." If a poison were bred and distributed there, all birds and beasts would be affected and die. Dekker was enraged by empirics, mountebanks, and quacksalvers, who, he said, are worse than the plague. "Only this Antidote apply, / Cease vexing heavn, and cease to die."

What then of the doctors who wrote about plague? Diemerbroeck said, according to the apothecary and later plague author William Boghurst, that "noe man had writt particularly of many things happening in the plague before hee did, being frighted with the Disease it selfe." He did not mean that the disease evoked too much horror, as though mere description of it might contaminate the writer, for the grim and even grotesque details of the disease were relished. He meant that the doctors who might have been in a position to understand plague as a medical condition were the first to flee from it when it appeared, so that "all their learning about it can be but opinionative and conjecturall."[10]

The theological motif was not always present. Some of the earliest printed plague treatises, those in French dating from the fifteenth century, bypassed supernatural explanation altogether to recognize both causes and signs of plague in nature. The signs and forewarnings might include strange winds, dark days, comets, an abundance of flies, and displacements of the stars; the causes might proceed either from above (venomous exhalations of the stars) or below (rotting corpses).[11] The doctrinal assertion that God was the primary cause of plague and that no secondary cause was operative was an extreme view that emerged later and was perhaps more pronounced in seventeenth-century English theology, in the Puritanism that placed every event under God's direct control. It is difficult to say whether the reawakened interest in secondary causes at midcentury owed more to the weakening grip of theology, to increased confidence in naturalistic explanations, or to the develop-

---

[9] Ibid., p. 60.

[10] Boghurst, *Loimographia*, p. 9.

[11] See, e.g., the incunabula reprinted in Klebs and Dorz, *Remèdes contre la peste*.

ment of a physico-theology in which the natural and the supernatural could be portrayed as complementary. By 1673, Dr. Thomas Burnet could repeat Diemerbroeck's notion that God, or rather "the anger of a most righteous God," was the primary cause of plague, but allow that there was also a secondary cause in a malignant, poisonous emission bred in the stars, and a tertiary cause in contagion from those already infected.[12] To primary, secondary, and tertiary causes, Burnet added disposing causes, such as evil conjunctions, hot and dirty air and skies, and "fetid exhalations of putrid things." But this array of causes, together with the recitation of forwarding signs and omens and the "marks" or "tokens" of plague that appear on its victims, constituted, rather than an explanation of plague, what might be called a semiology of plague, set in the semantic field of wrath, darkness, and poison.

The last major plague outbreak in Britain occurred in 1665, so physicians were for the most part deprived of the opportunity conveniently to study it in the last quarter of the century, with the result that treatises of the late sixteenth and early seventeenth century were reprinted rather than reconceived. As tempting as it might be to associate Diemerbroeck's ferments with Pasteur's, there is little real evolution in the concept. The notion that contagious febrile diseases corresponded somehow to ferments belonged to traditional medicine and had a phenomenological basis. The heat of decay and of fermentation suggests an analogy with the heat of fevers; dunghills steam, and, like corpses, may teem with worms. The multiplicative aspect of disease suggested something like the multiplicative action of yeast in bread or in a brew; a small quantity has the power to leaven a large amount and becomes itself the material that, in a small quantity, can leaven the next mass.[13] Fermentation and putrefaction, which were not clearly distinguished from other destructive processes such as rancidification and corrosion, seemed to be primary destructive processes that could result in the production of new living matter. They were not, until Pasteur studied "diseases of beer," regarded as secondary processes subsequent to infection by living organisms. If living organisms were identified or suspected to exist in buboes and festering sores, they were taken as effects rather than causes.

Girolamo Fracastoro, who had theorized about syphilis, typhus, and

---

[12] Burnet, *Thesaurus medicinae practicae*, p. 698.

[13] Thus Thomas Willis furnishes a well-received fermentational account of smallpox in his *De febribus*: A ferment is generated in the womb of the mother, which produces a swelling up and excretion of menstrual blood; this ferment is transferred to the child and, afterward, "being moved or stirred up by some evident cause, they ferment with the Blood and induce it to an ebullition, from whence many symptoms of the disease arise." *De febribus*, in *Diatribe duae medico-philosophicae*, chap. 15.

plague in the 1530s and 1540s, had noticed the close relationship between fermentation and putrefaction and had also argued that putrefying bodies emit invisible corpuscles. This may, he says, explain how
they are communicable, how one rotten apple infects the whole barrel.
In *On Contagion* (1546) he states that "all infections may be reduced
ultimately to putrefactions." But, as it turns out, contagion occurs in
other ways as well. Ophthalmia, he says, may be transferred without
contact, simply through one infected person's looking at another. And
transmission by *fomites*, by contact with the bedding, clothes, or uten-
sils of a sick person, though it involves "seeds of contagion," does not
appear to involve effluvia from putrefactions in the sick person's body,
for "those minute particles given off by a body affected with putrefaction do not appear to preserve their virulence for long and on that account are not to be regarded as of identical nature with those of *fomites*
or with those that act by contact alone."[14] In effect, Fracastoro was
simply restating the puzzling fact that diseases seem to be transmitted in
three ways: by direct contact with a sick person, by only apparently
harmless inanimate objects, cloth, wood, and so forth; and by action at
a distance originating with a sick person. But he does suggest that *semi-
naria* reproduce and that they have specific affinities for plants, animals,
and organs or fluids in the body.

If fermentation theories bear a certain somewhat misleading similarity to germ theories, so too do the accounts of Paracelsus and Jean
Baptiste van Helmont, which impressed such physicians as William
Harvey. Paracelsus had various categories for diseases: there were tar-
taric diseases, women's diseases, elemental diseases, and invisible diseases, the last arising from misuse of the imagination, emanations from
the dead, stellar emissions, and magic. Of interest in both authors' accounts is the conception of the disease state not simply as an abnormal
bodily condition but as a process of struggle between the patient, or the
patient's inner *archeus*, and an unwanted guest with destructive inten-
tions that has invaded the body from outside, or that is connatural.
"[A] disease is from an efficient seminal cause, positive, actual and real,
with a seed, Manner, species and order."[15] The illness-producing agent
was, on Helmont's account, an immaterial though "vital" parasite and
invader; he did not conceive it as another physical organism seeing a
home and a place to multiply. The notion they share is that disease
cannot proceed from a mere nothing—a privation, an imbalance in the
body of the patient—nor yet from some merely material process, but

---

[14] Fracastoro, "On Contagion" (1546), in Clendening, *Source Book of Medical History*,
p. 108.
[15] Helmont, *Oriatrike*, p. 805.

only from a real entity, "a strange guest received within and endowed with a more powerful or able *archeus*."[16] In some statements of the theory, the disease entity is conjured up by the patient's own *archeus*, so that disease is still, as in traditional medicine, essentially endogenous rather than inflicted. But the engendering takes place on the provocation of the *archeus* by some external irritant: for example, the saliva of the rabid dog, a plague "virus," the venom of the tarantula, or the juices of wolfsbane and nightshade. Disease is thus a conflict between two intentional and self-interested agents, the angry patient and the stubborn invader.[17] What does this explain? The burning and rage of infectious diseases, Helmont believed, their violence and paroxysms. He could not accept the view that disease was simply the result of a destructive physical agent; the Helmontian theory interprets even chemical poisoning vitalistically, an extreme that would be overcorrected by the corpuscularians. Presumably the struggle of the ill or poisoned person to survive is so dramatic and unpredictable that it could not be conceived otherwise than as directed against a living opponent.

The notion that disease is produced by an "idea" or a "form"—something immaterial that is able to replicate or stamp itself in the matter of the body many times over—and that fantasy and panic play a role in its transmission, as for Harvey imagination and desire play a role in conception, is the occult link between generation and contagion. When Harvey wrote to Giovanni Nardi in Florence that what especially delighted him in Nardi's book was "that part where I see you ascribe plague almost to the same efficient cause as I do animal generation,"[18] he was not supposing that seeds of plague that germinate in the human body were involved. Rather, he was pointing out that, in generation, an "idea, form or vital principle" is transferred from the genitor to the genetrix, and from her to the ovum and the fetus, which will produce the "image" of the species, but with the particularities of its parents and of its prenatal experience, and that something similar happens when a person is infected with a disease. The difficulty of explaining how this is possible was matched by the difficulty of explaining how plague and leprosy are communicated at a distance, how they may reside invisibly in cloth or furniture, or in the walls of a house. As the generative faculty produces many copies of itself, so the disease produces its like in another body, not in "one or two only, but in many, without respect of strength, sex, age, temperament, or mode of life, and with such violence that the evil can by no art be stayed or mitigated."[19]

[16] Pagel, *J. B. van Helmont*, p. 104.
[17] For further elaboration, see Pagel, *Aspects of van Helmont's Science*, pp. 38ff.
[18] Reprinted in Harvey, *Works of William Harvey*, p. 610.
[19] Ibid., p. 611.

Theories of animate contagion that qualify as true germ theories were not unknown before the introduction of the microscope, reminding us again that the ancients did not deny the existence of active *subvisibilia* and living agents of contagion, but did not dedicate themselves to their pursuit either. Marcus Terentius Varro in *De re rustica* notes that "precautions must be taken in the neighborhood of swamps ... because there are bred certain minute creatures which cannot be seen by the eyes, which float in the air and enter the body through the mouth and nose and these cause serious diseases."[20] Lucius Junius Moderatus Columella makes a similar, though not clearly suboptical, reference to "plagues of swimming and crawling things deprived of their winter moisture and infested with poison by the mud of decaying filth, from which are often contracted mysterious diseases whose causes are even beyond the understanding of physicians."[21] The seventeenth-century version of the theory had therefore the authority of these *prisci etiam gravissimi auctores* (ancient and respected authors), but, with the exception of Descartes—to whom Hobbes ascribed in passing the view of Varro—the moderns were hostile to the idea of a *contagium vivum*. They preferred to account for the replicative aspects of disease not by appeal to the replication of a form or a living entity, but simply by reference to the distribution of a toxic substance.

Boyle's confidence in micromechanical explanations for all macroscopic changes was particularly tested by the specific and rapid actions of poisons and drugs, which seemed irremediably "occult." Their toxic and beneficial effects alike, he argued, were due to the shape and motion—typically the sharpness or lubricity—of the particles composing them, not to specific occult qualities. Boyle wrote two essays of particular relevance to the question of the causes of disease, "On the Strange Subtilty of Effluviums"[22] and "An Experimental Discourse of Some Unheeded Causes of the Insalubrity and Salubrity of the Air."[23] The general theme of the first essay is that diseases may be caused by material particles of almost unimaginable smallness that mechanically disrupt bodily processes. Their ability to persist unchanged accounts for the long contagiousness of *fomites*, while their easy distribution should address the problem of epidemic diseases.

By way of persuading the reader that there are no purely incorporeal agents and that all active *subvisibilia* occupy space, Boyle began the treatise on effluvia by citing familiar examples of diffusion—cochineal

[20] Varro, *On Agriculture* 1.12.2.
[21] Columella, *On Agriculture* 1.5.6.
[22] Boyle, *Works* (1744 ed.), 3:661–706.
[23] Ibid. 5:38–70.

dye in water, small quantities of drugs in the body, the magnetic effluvia that pass through glass, the persistence or disappearance of smells. A pair of perfumed gloves kept its scent for thirty years lying in a drawer. Ambergris and asafetida do not lose weight when exposed to air, though their odor eventually weakens, while cloves and nutmeg do. But even the first observation, he said, does not prove that odors are immaterial, but only that the corpuscles released from odoriferous bodies are so small that the process of diffusion may continue for a long time without making any detectable difference in their weight. Those who deny this have too much trust, Boyle said, echoing Descartes, in the "negative information" of their senses. That such invisible effluvia may be harmful no one can deny; there are poisonous perfumes, and roses and musk cause headaches in some people, he said, even men, who are not subject to hysterical illnesses.

In his second essay, Boyle called attention to the diseases of miners, which were generally recognized as due to underground emissions of effluvia, and he noted that arsenic poisoning produces inflammations, pustules, and fevers like a contagious illness. He reviewed the alternative explanations of plague. Though the sacred writings taught that plagues have been inflicted in an extraordinary manner by God, we cannot, he thought, entirely accept this dogma. Animals, who are not eligible for divine punishment, also die in epidemics. Malevolent aspects of the stars, another hallowed cause, are "too remote, too general, too indeterminate," and too nonspecific to cause epidemics; and other putative general causes, such as the weather, seem vague and insufficient. The conclusion to draw was therefore that epidemic diseases are instances of widespread chemical poisoning, caused by exhalations from the earth, with orpiment the likely toxin. The corpuscular theory helped explain why chemical medicines might be useful as antidotes to poisoning and disease, preventing reactions in the body that would otherwise occur, as when saltpeter corrodes silver but not if you add salt to it. It also explained why, because of the antitoxic properties of the soil, some countries such as Ireland could not bring forth poisonous animals. Healthy air would also disable plague particles in the air and check the activity of "morbifick ferments."

The reference to "ferments" indicates that it is not the case that these corpuscles are inert, having nothing more than a disposition to cause abrasive damage; they may also, Boyle allowed, be self-multiplying:

And ... I think it very possible, that diverse subterraneal bodies, that admit effluvia, may have in them a kind of propagative, or self-multiplying power (to explain why diseases should remain endemic). I will not here examine whether this proceeds from some seminal principle, which many chemists and

others ascribe to metals and even to stones, or . . . to something analogous to
a ferment, such as in vegetables enables a little sour dough to extend itself
through the whole mass; or such as, when an apple or pear is bruised, makes
the putrefied part, by degrees to transmute the sound into its own likeness.[24]

These seeds may lie dormant for many years, as in the documented
example of the contagion preserved in a cobweb; or the pestilence kept
folded up in a linen cloth for fourteen years, which then broke out
again, reported by the atomist Daniel Sennert; or the carbuncle that
appeared on Diemerbroeck's foot after he touched with it some straw
that had lain under the pallet supporting the bed where a servant had
been sick of the plague eight months earlier. It is thus difficult to say
whether, for Boyle, the causes of plague were "living." Not without
reason has it been said that the late seventeenth century had no determinate concept of life,[25] and corpuscularian philosophers enjoyed considerable latitude and took advantage of considerable vagueness in imagining the powers, behavior, and effects of their corpuscles: one might say
that the seventeenth century had no determinate concept of mechanism
available to it either. When Boyle says that plague is "carried on mainly
by a malignant disposition in the air; without which some plagues could
never have been so catching, as they were," he leaves it unclear whether
the plague toxins have altered the very properties of the air so as to
render it poisonous or are simply present interspersed among the air
corpuscles. The important question for Boyle is really that of ultimate
origins: plague is no innovation, he wants to show, but part of nature's
storehouse; it comes from below, not from above, and this is also the
position of the London apothecary Boghurst, who wrote (but did not
publish) in 1666 *Loimographia*, a slim volume describing plague's
"Generation, Progresse, forerunning . . . Common Signes, Good and
Evill Signes, Meanes of Preservation, Method of Cure . . . with a Collection of Choice and Tried Medicines."

Boghurst decisively rejected the theological and endogenous interpretations. Plague did not come from God, nor did it come from within
man. It was no different in principle, he stated, from other diseases,
such as rabies, leprosy, smallpox, and French pox, for which no supernatural account was sought. "Must we ascribe every effect of the cause
whereof wee are ignorant to some cause produced *de novo* and immediately from God?" When God sends a judgment he always employs
natural means to do so—for example, assembling insects, or amassing
clouds—"not by any new-created Agents, but by the ministry of known
and second Causes, making them the Executioners of his Decree upon

[24] Ibid., p. 46.
[25] As Foucault maintained in *Order of Things*, chap. 8.

Mankind."[26] He also rejects the traditional views that the disease involves the excess of a manifest quality such as cold, heat, moisture, or dryness, or that it is the result of an occult quality such as a venom; he argues that the consensus is that it is not a putrefaction deriving from a manifest quality either. The effects of the disease, he says, are "soe strange and horrid" that manifest qualities are excluded as causes; plague is not just, not essentially, a fever, for all remedies against putrefying fevers are useless against plague. Boghurst sides with the moderns.

> That this venome is a body or a concretion of many little bodyes, though very subtle and invisible, can bee noe doubt for those that have outgrown Aristotle and are acquainted with the Epicurean or corpuscular philosophy; and for the nature of the pestiferous corpuscles and the manner of the propagation in the Air and humours of the body, light may bee had as from the Epicurean Hypothesis in generall, soe particularly from Lucretius in the end of his 6th booke and Gassendus. . . . But for the shape and figure of these Atomes or small bodyes, which is the foundation of their activity and of that power whereby they corrupt the texture, alter and change the motion of these corpuscles or particles which compose the spirits and blood, wee can say nothing to satisfaction; they fall not under the eyes' perception, though assisted with the best inventions in perspective wherein this last age hath furnished us with, or any other sense, and therefore wee may conclude with the words of Fernelius (*De abditis rerum causis*, lib. 2 cap. 12) "The seedes of the Pestilence are soe hidden and removed from sense that wee see them better in their effects than we can in themselves."[27]

Having established that the causes of plague are invisible and hidden, that they are atoms whose shape and figure gives them the power to damage or destroy bodily fluids, Boghurst turns to the problem of their origin. He rejects the hypothesis that they are generated in the heavens or stars, noting that this has been the "catholique and received opinion," on the ground that he does not understand how this could happen. He adds that, according to the Cartesian vortex hypothesis, fumes given off from comets might pollute and contaminate the air; this, he says, is more plausible than the notion that the stars and planets emit poisons.

The second proposal he rejects is that plague particles have an internal origin in a corruption of humors, which then spreads from one body to another. Boghurst does not deny that corrosive poisons can be produced within the body, and he appears to endorse the very old view that smallpox is a connatural disease acquired in utero from the mother's

[26] Boghurst, *Loimographia*, p. 16.
[27] Ibid., p. 10.

menstrual blood. Nor does he deny that some occurrences of plague might have such an origin, but he does not see how the body's humors can produce anything so malignant or epidemic. The fault, he decides, lies in the earth, "the seminary and seedplott of these venomous vapours and pestiferous effluvia which vitiate and corrupt the Aire." Plague is, in summary, "a most subtle, peculiar, insinuating, venomous, deleterious Exhalation arising from the maturation of the ferment of the Foeces of the Earth extracted into the Aire by the heat of the sun, and difflated from place to place by the winds, and most tymes gradually but sometymes immediately aggressing apt bodyes."[28]

The evidence Boghurst cites is similar to Boyle's: poisonous minerals in the earth, the short lives of miners, the paroxysms and convulsions of Paracelsus's servant who breathed noxious vapors, the "mortiferous and venomous vegetables" that grow in the earth, which are but "concretions or grosser consistencies of those virulent Particles." Later in the treatise he observes that the contraction of the jaws of plague victims has a "sympathetical correspondence" to the effects of mercury poisoning, suggesting that the toxin of one is similar to that of the other.[29] Venomous animals live on the earth too, and, dead or alive, may infect the air by their exhalations. Earthquakes tend to produce plagues; the release of vapors from an opened tomb caused a plague, as historians recount, in Roman Silicia.

Diemerbroeck is wrong, Boghurst argues, to think that nothing so evil can be among the products of creation, for a full inventory of "all the poysons in Nature's Storehouse, whether animall, vegetable, or minerall," shows a wonderfully potent armory. "Good" refers, as one might learn from every common writer of metaphysics, not to what is pleasant but to what is suited to an end. Venomous and hurtful creatures are thus good things insofar as they enable people to exercise their reason and prudence, those God-given characteristics that would have been useless to them otherwise.

Noteworthy in Boghurst is his insistence that we must attend to the "quiddity" of the disease, what it is in itself, not its effects or circumstances. So he rejects the environmental and contextual itemizations of folk medicine and theory alike. Close dwellings, poor diet, uncleanliness, dunghills, dead bodies lying unburied, unsuitable weather, an increase of vermin, too much venery, rotten meat and moldy bread, eclipses, and so on may exacerbate but cannot cause plague. "Good aires and sweet and cleanly places were no more exempt from the disease than stinking places, and healthfull wholesome bodyes fell under

[28] Ibid., p. 19.
[29] Ibid., p. 40.

the disease as much as pocky corrupt bodyes, and more too." In the end Boghurst cannot resist drawing a moral. Though plagues always come unpredictably, they are always just; wherever they come "wee shall bee ready to make way for them by our sinnes and deserts."

There was still a contest in 1670 among act-of-God theories of the origins of plague, stellar and planetary accounts, environmental accounts, and corpuscular theories, which represented plague as an accidental occurrence produced by the unfortunate interaction of human bodies with plague particles following geophysical upsets. Boyle considered that the multiplicative faculty of the disease implied some kind of analogy to organized life or to the aggressive processes of fermentation and putrefaction. He did not attempt to adjust his philosophical mechanism to deal with this paradox—or with the problem of inanimate self-replication—but simply left open the suggestion that diseases might be or contain "seminal principles." And this was still the view of Giovanni Maria Lancisi in 1715: plague is a ferment, involving the proliferation of corpuscles that irritate, corrode, and damage the animal machine.[30]

Mention of plague "seeds" here, as elsewhere, does not imply a specific theory of animate contagion, a development signaled rather by the introduction of references to "eggs," "ova," "ovae," or "ovula." Gabriello Fallopio in *De morbo gallico* (1564) said that living, exhaled blood corpuscles produced phthisis and syphilis when they were breathed in by a healthy person and fell, as it were, on suitable soil, infecting it and rendering that person contagious in turn.[31] Diemerbroeck too described plague as caused by "secret, malignant, and virulent Seeds; which consisting of very subtle and volatile particles, whensoever they are put into a Commotion, are effectively dilated, and disperse their pernicious Contagion throughout the spacious atmosphere" in the manner of some perfumes.[32] But the perception that diseases begin invisibly and grow and that they may be "sown" here and there before they spring up still leaves out the critical reference to the life cycle of a specific organism.

At least one common contagious human disease was known since ancient times to be caused by minute insects not clearly visible to the naked eye. That disease was scabies, which is caused by a mite, *sarcoptes scabei*, and results in blisters and itching. Andry notes the priority of the Arabs, in particular the physician Avenzoar, in describing the disease in his treatise on internal parasites, and he even raises the question whether other Arab physicians might have been acquainted with

[30] Lancisi, *De bovilla peste*, chap. 8. Lancisi leaves it open whether the process involves microscopical animals or an inanimate miasma.

[31] Singer and Singer, "Doctrine of *Contagium Vivum*," pp. 189–90.

[32] Diemerbroeck, *Treatise concerning the Pestilence*, p. 3.

and used microscopes; he says correctly that Aristotle was familiar with small cutaneous insects as well.[33] Since ancient times, these "hand-worms" (the term "worm" refers to creeping or crawling animals gener-ally) had been extracted from the skin with the tip of a needle, and various powders and ointments were tried against them with at best temporary results. It was apparently the Italian Giovanni Bonomo who first examined the scabies mite microscopically, finding it to be "a very minute *Living Creature*, in shape resembling a Tortoise, of a whitish color."[34] Bonomo saw the creature's eggs, and concluded that the mite could be spread by contact with sheets, towels, handkerchiefs, or gloves previously used by itchy persons, by transmission of either animals or eggs. The disease could not be cured, he thought, by internal medicines, but only by applying strong baths, washes, and ointments, and a per-manent cure was not to be expected because of the ease of reinfection. In his *Osservazioni intorno a' pellicelli del corpo umano* (1687), he ar-gued against Aristotle and his many followers on this issue that para-sites were not products of the diseased human body but causes of dis-ease; elimination of the parasite brought healing, and transmission of the parasite brought new symptoms. This claim to have observed a dis-ease-causing organism of a recognizable type and to have sketched a mode of transmission distinguishes animalculist theories of contagion from the vaguer ideas of contagious living effluvia proposed by earlier writers. According to the Singers, Victor Bonagentibus in 1556 had compared the generation and conveyance of fevers to the putrefactive processes that produce worms in corpses.[35] But such a comparison sim-ply reflects the old observation that fermentation is a source of heat and that fermentation and worm-producing putrefaction are allied pro-cesses.

A different understanding emerged in the thirty years between 1650 and 1680, when the increasing use of microscopes led to several state-ments of the doctrine of contagion by animalcula or insects. Borel found minute insects and their eggs in cases of virulent gonorrhea (this was the observation that perhaps inspired Ham to look for them him-self), and worms in the pores of the skin and in smallpox lesions, and he observed in this connection that the encyclopedist J. H. Alsted had claimed that in times of plague the ambient air is full of invisible ani-

---

[33] Andry, *Breeding of Worms*, p. 49. We find the following in Aristotle, *History of Animals* 556b: "The slightest quantity of putrefying matter gives rise to fleas; bugs are produced out of the moisture from living animals as it congeals outside them, lice are produced out of flesh. When lice are going to be produced, as it were small eruptions form, but without any purulent matter in them, and if these are pricked, lice emerge."

[34] Bonomo, "Concerning the Worms of Humane Bodies."

[35] Singer and Singer, "Doctrine of *Contagium Vivum*," p. 189.

malcula arising out of its corruption.[36] But the elevation of the animalculist theory to prominence is due to August Hauptmann. In a pam-
phlet of twenty-three pages published in 1650, Hauptmann praised the
microscope for its revelation of the secrets of nature and stated that
fevers are caused by minute, spontaneously generating insects: "[V]ery
minute and almost invisible animalcules are the cause of all deaths in
men and animals. The creatures are minute wormlets beyond the reach
of the unaided senses. They form their own tiny bodies from certain
kinds of very subtle corruption." There is still some uncertainty whether
these animals are products or causes or both of disease: "[A]nimalcules
(*cridones sive Dracunculi*), insects, moths, and little corroding malignant worms and acari swarm in the discharges and humours of measles,
scabies, and small-pox, and both give rise to the lesions and cause their
rupture."[37] But the idea is that death is caused by the act of a living,
alien, and hostile organism "welches nichts anders als eine Wurm-
maessige substanz ist" (resembling a worm), and which we carry with
us as a "lebendiger Tod"—a living death. The paradoxical nature of
disease as self and not-self, as dead and living, was on its way to becoming resolved.

Hauptmann claimed that, risking scorn and mockery for his views, he
had sent his *Epistola praeliminaris* to the Roman savant Athanasius
Kircher. He reprinted it with Kircher's reply in his treatise on mineral
springs published in 1657, and Kircher presented his own animalculist
theory in his *Scrutinium physico-medicum* (1658). Here he argues that
all rotting substances are full of "worms" that propagate infection. Plague
is not simply an internal process involving corruption of the bodily humors, nor does it come from outside in immaterial form like the unin-
vited guest of Helmont. It is an alien, traveling, deleterious faculty, but
one with a material embodiment. According to Kircher, this subvisible
effluvia animata is the cause of plague, as well as leprosy, venereal diseases, and elephantiasis: "Air, water and earth teem with innumerable
insects capable of ocular demonstration. Everyone knows that decom-
posing bodies breed worms, but only since the wonderful discovery of
the smicroscope [*sic*] has it been known that every putrid body swarms
with innumerable vermicules, a statement which I should not have believed had I not tested its truth by experiments during many years."[38]
Issues of piety and naturalism are raised: the knowledge that plagues
come from God and are punishments for the wicked world does not
prevent Kircher from asserting that plague also has a natural cause, and

[36] Borel, *Centuria observationum*, p. 37.

[37] Hauptmann, *Epistola praeliminaris, tractatui de viva mortis imagine*, in Singer and
Singer, "Doctrine of *Contagium Vivum*," pp. 196–97.

[38] Kircher, *Scrutinium*, in Singer, "Early History of Microscopy," p. 278.

that there are certain preventive measures one can take and that it is
morally justifiable to take. Though piety, repentance, and acts of mercy
are the most significant steps one can take against plague, they do not
preclude flight from the city, good ventilation, and the use of odorous
substances.

Kircher regarded his investigations as confirming the general produc-
tivity of nature; serpents, dragons, and toads may be generated in
mountainous caverns and caves; plants generate worms from their sap,
as do all cadavers from their flesh. Microscopical worms in turn arise
from the corruption of humors in the body, chemical reactions, bad
food, poisonous plants, and the breath of malignant animals, serpents,
toads, newts, and rabid dogs and cats. The experiments he describes in
this connection involve preparing infusions of rotten meat, especially
snake meat, and exposing them to the light of the moon:

> Take a piece of meat, and at night leave it exposed to the lunar moisture until
> dawn of the following day. Then examine it carefully with the microscope
> and you will find that all the putridity drawn from the moon has been trans-
> formed into numberless little worms of different sizes, which in the absence of
> the microscope you will be unable to detect, no matter how sharp-sighted
> [you are], with the exception of those which have grown to such a size as to
> become visible. You will have the same experience with cheese, milk, vinegar
> and similar substances undergoing putrefaction. However, you must not sup-
> pose the microscope to be an ordinary one; but highly wrought by a careful
> and practised hand; such as mine is which represents objects a thousand times
> larger than they really are.[39]

The secondary literature has been critical both of Kircher's experimental style and of the relevance of his findings. Although his text describes
the generation of worms from putrefying matter with the help of the
moon and seems to furnish no direct evidence that these worms are the
causes of disease, Kircher does argue that the seeds of these worms are
carried through the air and lodge in the body, where they produce a
putridity of humors. Thus, while stating it as a principle that "omne
putridum ex se et sua natura vermes generat" (every rotting substance
generates worms by itself through its own nature), he does advance in
two ways beyond the traditional idea that putrefaction produces insects.
First, he claims that these insects are not generated directly from rotting
matter but from seeds, ruling out metamorphosis; second, he indicates
that putrefaction is the product, not simply a contributing cause, of the
life processes of microscopical organisms.[40] All substances in nature, he

---

[39] Kircher, *Scrutinium*, in Torrey, "Athanasius Kircher," p. 257.
[40] Kircher, *Scrutinium*, p. 41. But cf. Torrey, "Athanasius Kircher," p. 265, for a negative assessment. Kircher's microscope, which, he says, magnified objects one thousand

says, "exhale certain effluvia composed of extremely minute invisible corpuscles." These corpuscles may be fragrant, refreshing, and beneficial to the body, or oppressive and destructive. "Corpuscles of this kind are commonly without life but through the agency of the circumambient heat already tainted with a similar pollution, they are transformed into a brood of countless invisible worms. . . . So that the effluvia may now be called, not lifeless but animate."[41] Because the corpuscles are exceedingly light,

> they are agitated like atoms by the least breath of air; but since they have a certain viscosity and glutinous tenacity, they insinuate themselves very easily into the inmost fibres of clothing, cords, and linen stuffs; in fact by virtue of their subtlety, they penetrate any porous material such as wood, bone, cork, nay even metals; and there they bring forth new *seminaria* of contagion; and inasmuch as their substance is extremely tenuous, they live for a very long time merely on the moisture which comes to them from without.[42]

Kircher also told Monconys in 1664 that he had found the buboes of plague victims and the pimples of persons with "copperrosée" full of microscopical worms, which filled the air in times of pestilence.[43]

The Royal Society showed flashes of interest in the possible role of animalcula in disease. John Wilkins reported in February 1663 that William Croone had seen minute insects in the blood of a dog he had dissected; Croone, a relatively careless observer, later admitted that the minute insects had been spotted only on the knife he had used, and that after a period of two weeks. Mr. Hoskyns of the society was appointed to look into the status of Kircher's theory, but this seems not to have happened; many Royal Society assignments were not fulfilled, or at least not reported on. In 1664, Dr. Merrick observed that Harvey had found the blood of febrile patients full of worms. And in the memoirs of the Royal Society for 1665–66 there is a notice by Walter Charleton saying that the theory of the vermination of the air, which originated with George Ent and had been furthered in Italy by Kircher, was then well advanced in Italy, it having been observed in Rome that "there was a kind of insect in the air, which being put upon a man's hand, would lay eggs hardly discernible without a microscope; which eggs, being for an experiment given to be snuffed up by a dog, the dog fell into a distemper accompanied with all the symptoms of the plague."[44] Henry Power, after observing "pond mites" with a microscope, speculates in

---

times (i.e., 10 x 10 x 10) would not have been sufficient to see bacteria in any case; what the "worms" he saw in blood were remains doubtful.

[41] Kircher, *Scrutinium*, in Torrey, "Athanasius Kircher," p. 269.

[42] Ibid., p. 270.

[43] Monconys, *Journal des voyages* 2:445.

[44] Birch, *History of the Royal Society* 2:69.

Experimental Philosophy that "not onely the water, but the very Air it self, may certainly at some times and seasons be full of Living creatures; which must be, most probably, when great putrefactions reign therein, as in the Plague-time especially." With his telescope he thought he could see, on some days, "a tremendous Motion and Agitation of rowling fumes and strong Atoms in the air."[45]

An "observing person in the Country" who wrote the editor to comment on the all-around excellence of *Philosophical Transactions* 133 speculated, without, he said, being able to prove it, that disease was associated with the infestation of the air by animalcula: "Mr. Leeuwenhoek's Microscopical Discoveries are so exceedingly curious, and may prompt us to suspect, that our Air is also vermiculated, and perhaps most of all in long Calms, long-lasting Eastern Winds, or much moisture in the Spring-time, and in seasons of general Infections of Men or Animals."[46] The writer did not really say whether the vermiculation of the air was the cause, effect, or both of the disease, but he did suggest that breaking up stagnant air with music, guns, bells, and shouts, and cleaning it with fire and water, might rid it of dangerous animalcula, as smoke destroys caterpillars and other pests in orchards and gardens. The German savant J. C. Sturm said in 1687 that the air was full of homunculi and animalcula that were breathed into the lungs and that, unless they were sweated out again, could excite diseases.[47] The preface to Christian Lange's *Pathologia animata* (1688) shows a physician scrutinizing the air outside a hospital with a tubular microscope.

Since the time of Hippocrates, "bad air" had been recognized as a cause of disease; the theory of animate contagion, which here makes its appearance as a theory of "vermiculation," required that the bad agent in the air be understood as living and endeavoring to persist in its being, rather than as dead, foul, or noxious. Despite some ambiguity, this seems to be Kircher's picture and also that of Frederick Slare, who, in the *Philosophical Transactions* for 1683, relayed a report of a "blue mist" that seemed to propagate a cattle epidemic in Switzerland; he suggested that it was "worth considering whether this infection is not carried on by some volatile insect, that is able to make only . . . short flights" and expressed regret that "Mr. Leeuwenhoeck" had not been present at the dissections of the sick animals.

Though no medical writer claimed personally to have seen microscopical worms in the air, worms were detected in all manner of bodily or-

[45] Power, *Experimental Philosophy*, p. 115.
[46] "Some Considerations," p. 891.
[47] Buetschli, "Infusoria," part 3, p. 1107.

gans and fluids.[48] After Borel and Kircher, Sieur le Guide said, according to the Singers, in a 1676 treatise on venereal diseases, that he had observed with the microscope a kind of insect in luetic ulcers that had "the appearance and slow movement of snails," which explained, he thought, why syphilis was transmitted only by contact rather than through the air.[49] Yet, as the Singers remark, he regarded these insects as the effects of the disease, produced by the putrescence of the ulcer: the agent of contagion was not necessarily identified with the cause of tissue destruction or the other symptoms of the disease. Christian Franz Paullini, in his report on dog shapes, *Cynographia curiosa* (1685), finds worms "everywhere in the Microcosmos," and in every element including fire, though they are visible only to the eye armored with a lens. They are to be seen in saliva, tears, the nose, fruits, plants, vegetables, and meats; they cause phthisis, puerperal fever, sweating sickness, measles, smallpox, cancer, headache, bewitchment, and convulsions.[50] The worm, he says, is the instrument of God's anger; what we call natural death is only the process of being devoured by it. According to a French review of the book, "It is known . . . that air and food are often full of an infinity of insects and of thousands of invisible seeds which we devour, and one might suppose that these insects and seeds find in our bodies matrices where they are rendered fertile by our natural heat."[51]

The notion that natural death is destruction by invisible worms drew support from two sources: first, medical experience with macroscopic worms, such as the roundworm, *ascarides lumbrici*, and the tapeworm, which were known to infest the guts of humans and their livestock; second, insight into the true prevalence of parasitism. The size, number, location, and figure of worms voided by adults and children or found in bodies after autopsy were regular features of interest in the reports of the scientific academies; Kerckring said that he had even found parasitic worms in a six-and-a-half-month-old fetus.[52] The discovery of insect and protozoal microparasites was furthered by Leeuwenhoek, by Redi, and by Vallisnieri. Redi found that every creature he examined, with the exception of tigers and leopards in menageries, was infested with a distinct species of lice; even frogs, ants, and beetles had their parasites; even parasites had their parasites.

Nicolas Andry's book, *De la génération des vers dans le corps de l'homme*, notable for its defense of the role of spermatic worms in gen-

[48] See Belloni, "Micrografia illusoria e 'animalcula.'"

[49] As reported in Singer and Singer, "Doctrine of *Contagium vivum*," p. 201.

[50] Paullini, *Cynographia curiosa*, pp. 133–34; see also Singer and Singer, "Doctrine of *Contagium vivum*," p. 210.

[51] *Nouvelles de la république des lettres* 1 (November 1684): 165.

[52] According to Andry, *Breeding of Worms*, p. 79.

eration, was printed with a seal of endorsement from the medical faculty of Paris in 1700 and went through many editions, including an immediate English translation. The title suggests, wrongly, that Andry subscribed to the time-honored view that men generate their parasites. He began his treatise by observing that the microscope has shown that insects are not "rough Drafts or imperfect Productions of Nature" but fully formed animals with lungs and hearts: this is as true of the smaller insects as it is of the larger (snakes, frogs, etc.). Even the most minute of those animals that are found in the bodies of the larger animals are bred from seeds: "If we consider the Eggs of Caterpillars, Flies, and other small Insects with the almost infinite Number of little Animals, which Microscopes discover to us, we shall easily find that there is nothing in Nature, into which the Seed of Insects may not insinuate it self, and that a great Quantity of them may enter into the Body of a Man, as well as into those of other Animals, by means of the Air and Ailments."[53] Worms in the gut may cause vomiting, pain, bloating, sweating, convulsions, stagger, and muteness; microscopical worms may wreak even worse damage. Smallpox, cancer, and venereal disease are probably all worm diseases. Tumors and cancers are in all likelihood produced when minute insects chew through the sieves of a gland so that it no longer absorbs selectively but swells to a great size; as the insects continue to eat, ulcers are produced. He draws on Swammerdam's accounts of the metamorphosis of many insects to argue that worms in the body may assume the forms of other creatures:

> The worms that breed in Humane Bodies, whether within or without the Guts, do oftentimes assume monstrous Figures as they grow old; some take up the Shape of Frogs, others of Scorpions, and others of Lizards. Some shoot forth Horns, others acquire a forked Tail; some assume Bills, like Fowls, others are covered with Hair, or become all over rough; and others again are covered with Scales and resemble Serpents.[54]

Spermatic worms, Andry said, need to be distinguished, as Nicolaas Hartsoeker had done, from the flesh-eating variety. The former, he agreed, are not really worms, as they are not found in children, venereal patients, or old men. Hartsoeker was entirely in agreement with him at the time on the role of both types; he thought that sick people had in their bodies "an infinite number of invisible *Insects*, who grow and devour everything that comes in their way."[55]

But skepticism about these theories ran high from the beginning. Die-

[53] Ibid., p. 9.
[54] Ibid., pp. 80–81.
[55] Letter to Andry, 11 June 1699, in ibid., pp. 216–17.

merbroeck, the great authority on plague, had already considered and rejected any form of animate contagion involving "invisible animalcules" in 1665,[56] and Kircher's theory is discussed and ruled out in the 1672 edition of Nathaniel Hodges's *Loimologia*. Hodges considers whether or not in putrid fevers worms might be found in the lesions of sick patients, as Kircher had suggested. He says that no such insects have been detected and suggests sarcastically that it is perhaps only the Italians, with their better weather and clearer skies, who are able to see them with their lynx eyes. He himself advances a "nitro-aereal" theory, according to which deleterious particles, airborne but of terrestrial origin, are the cause.[57]

Interest in the theory was carried forward nevertheless into the first decades of the eighteenth century. An epidemic among cattle, introduced into Venice in 1711 and reaching a climax in 1714, prompted renewed speculations about a *contagium vivum*. It was Vallisnieri, propping himself on the authority of Kircher, Hauptmann, Borel, and Lange, who reintroduced the idea of contagion through minute worms that have appropriated the circulatory system and penetrated deep into the body. He suggests that the mode of treatment might be a simple, though still unknown, poison, "perhaps growing in the gardens of the poor," fatal to these worms that could be infused into the veins to exterminate them.[58] But Vallisnieri remained in some doubt about the theory and challenged Andry aggressively for his poor scholarship, his faulty logic, and his credulity. An outbreak of plague in Marseilles in 1720 again occupied the attention of French and Italian savants, including J. B. Goiffon, the author of the *Relations et dissertation sur la peste* of 1722, and two Englishmen who argued passionately for the animalculist theory, Richard Bradley, a botanist, and Benjamin Marten, a physician.

Part 3 of Bradley's *New Improvements of Planting and Gardening* (1718) included a chapter on blights and insect infestations, which he had observed to affect plants selectively according to species. He reviewed the contributions of Hooke and Leeuwenhoek to microscopy and cited the discovery of vast numbers of animalcula in a single drop of water, some of them eight million times smaller than a grain of sand,

---

[56] Diemerbroeck, *Tractatus de peste*, p. 48.

[57] Hodges, *Loimologia sive pestis nuperae . . . narratio*. The handwritten margin notes in my copy say "We have not yet detected any insects. . . . I believe the microscope will ultimately afford the most important information." The writer (from the nineteenth century?) is referring to a cholera outbreak.

[58] Antonio Vallisnieri, letter to C. F. Congrossi, in Belloni, "Charlatans," p. 580.

with muscles, organs, life, and motion. Bradley reproduced with approval the pananimist sentiments of a correspondent. No part of earth or water is free of the seeds of life: "Nothing is out of Motion or can be void of it."[59] He agreed that vapor sicknesses and illnesses brought on by smells are utterly different from contagious diseases: "The most nauseous *Vapour* of it self will not cause any *Distemper* that is *Epidemical*."[60] Transmission and infection is typically by air. "There are Beings subsisting, which are not commonly visible, and therefore it is easie to conceive that the most gentle Air, is capable of blowing them from Place to Place."[61] The common people, he notes, share his opinion that agricultural blights are correlated with, but not brought directly by, the east wind. They break out, he suggests, when the wind hatches the eggs of certain caterpillars; in response farmers burn heaps of weeds and chaff to poison the eggs and caterpillars or dust them with pepper powder.[62] Sometimes the country people are deceived: the London farmers believed that the birds that descended one year on their turnip crops had eaten the leaves, until Bradley showed them that the birds were friends rather than enemies and preyed on the destructive caterpillars.[63]

Bradley states clearly that plague is not sui generis, but one of the numerous blights and epidemics caused by another living species that affect the plant and animal worlds. He takes up the problem of human epidemics in *The Plague at Marseilles Considered* (1721). After describing Marseilles, its large number of churches and religious institutions, its harbor location, and its countryside, Bradley tells us that the city was as backward in the year of its plague as London had been during the last great outbreak of 1665: the air was bad, the streets were narrow, the diet was mostly vegetarian. That no plague had since occurred in London was due, he suggests, to the purifying effects of the Great Fire of 1666, which, "besides Destroying the Eggs, or Seeds, of those Poisonous Animals, that were then in the Stagnating Air, might likewise purifie the Air in such a Manner, as to make it unfit for the Nourishment of others of the same Kind, which were Swimming or Driving in the Circumambient Air." The larger and cleaner streets that followed the rebuilding also cleansed the air of sustenance for animalcula.

Appending to this preliminary discussion a long sequence of mortality tables, Bradley notes that the plague was at its height during the warmest months of the year, July through October, and that in the summer months one can often observe tiny insects floating in the air. Again he

[59] Letter from Robert Ball to Mr. Bradley, in Bradley, *New Improvements*, p. 88.
[60] Ibid., p. 86.
[61] Bradley, *New Improvements*, p. 66.
[62] Ibid., p. 57.
[63] Ibid., p. 54.

recalls that tree blights affect certain species—for example, cherry trees—selectively; other diseases befall only plants whose leaves have already been injured, for example, by hailstones. We must suppose that each type of insect "has its proper *Nidus* to hatch and perfect it self in; and is led thither by certain Effluvia which arise from that Body which is in a right State for the preservation of it." And, just as plants attract their parasites to them, "so in Animals it may be, that by ill Diet the Habit of their Body, may be so altered, that their very Breath may entice those poisonous Insects to follow their way, till they can lodge themselves in the Stomach of the Animal, and thereby occasion Death."[64] Finding their nidus, these insects "will certainly lay their Eggs there, which the Breath of the diseased Person will fling out in Parcels, as he has occasion to Respire; so that the Infection may be communicated to a stander by, or else, through their extraordinary smallness, may be convey'd by the Air to some Distance."[65]

Disputing the view that "Noxious Vapours" may be responsible for epidemics, Bradley describes another blight that "poisoned and destroyed" two hundred miles of trees and that was associated with worms that broke out of holes in the ground and turned into flies, which stuck their stingers into the trees and killed them. Moving to the animal kingdom, he discusses the celebrated "blue mist" cattle plagues in Switzerland in 1682, which some said were caused by the witchcraft of Capucin monks, but which "the more learned believe . . . to proceed from some Noxious Exhalations thrown out of the Earth by three distinct Earthquakes." Again it is worth considering, he thinks, echoing Slare, "whether this Infection is not carried by some volatile Insect" as a hypothesis preferable to that of the ancients, who derive disease from "a blind Putrefaction, from the incantations of ill Men, or from the conjunction of inauspicious Planets." Other plagues from Italy and Great Britain are discussed before Bradley returns to his theme of insect infestation. The worms in men, women, and children, he thinks, "cannot breed in such Bodies from nothing, without either their Eggs or themselves are brought thither by some Accident: For if they were the natural Produce of Animal Bodies, they would then be alike common to all, which we know they are not." They must therefore be "suck'd into the Stomach with the Breath, or taken into it with some unwholesome Food."[66] That there was no plague in America supported his view that the pestiferous insects are blown hither and thither by winds across Asia and Europe but cannot cross great tracts of ocean. Bradley does not

[64] Bradley, *Plague at Marseilles Considered*, p. 21.
[65] Ibid.
[66] Ibid., p. 36.

pretend to be discussing all diseases of men, but his formulation of a theory of epidemical disease is sweeping. "[A]ll Pestilential Distempers," he says, "whether in Animals or Plants, are occasion'd by poisonous Insects convey'd from Place to Place by the Air, and by uncleanly living and poor Diet, Human and other Bodies are disposed to receive such *Insects* into the Stomach and most noble Parts."[67]

Bradley's arguments are purely circumstantial and analogical. He in no case claims to have seen or otherwise identified a disease-producing insect in the human body, nor does he refer again to the revelations of the microscope until near the end of his treatise. He does eventually tell us that the dead bodies of the Marseilles victims were found full of insects "and that those Worms could be no way so suddenly killed, as by putting Oil or Lemon Juice upon them," suggesting that they were no ordinary products of putrefaction. But he thinks that the salutary effect of liquors and tobacco is due to their toxic qualities: "I suppose therefore that the Smoak of Tobacco is noxious to these Venomous Insects, which I believe to be the Cause of the Plague, either by mixing itself with the Air and there destroying them, or else by provoking the Stomach to discharge itself of those Morbid Juices which would nourish and encourage them." The apparent success of some aromatics in warding off plague in 1665 is further evidence for the theory. As strong-smelling herbs are not normally attacked by insects, so rue, garlic, and other potent herbs, which are the basis of antiplague folk medicine, repel or kill the insects responsible.

Bradley based his case on similar patterns in plant and animal infections; Benjamin Marten, in *A New Theory of Consumptions* (1720), turns directly to the microscope. As a physician he was familiar with every medical theory of the causes of consumption, and in his survey of opinions he includes Paracelsus's view that consumption results from the corroding action of a tartar on the lungs, Helmont's theory of a ferment and obstructing mucilage, Sylvius's view that consumption follows an oversupply of blood to the lungs, which eventually causes irritation and ulcers, and Willis's sour-blood or defective-nervous-juice theory. To these theories and their variations, he appends a whole list of what are traditionally believed to be predisposing causes: intemperance, lack of exercise, unusual discharges, bad air, heredity, too much sleep, infection, chalky stones, and other diseases such as scurvy, king's evil, and gout. But all of these explanations, Marten says, leave us in the dark about the cause of consumption. The essential viciousness of the saltiness, sharpness, sourness, or ill nature of the humors found in tuberculous lesions is not explained by any of the medical writers; what is

---

[67] Ibid., p. 57.

malignant and deadly in the lesions remains mysterious. There is, he shows, a tautological aspect to these supposed reductions of the disease process to an underlying physical or chemical microprocess; all we learn from these accounts is that deleterious qualities have deleterious effects. Acid-alkali theories of disease and mechanical theories propounded by would-be Newtonians are equally useless.

Marten urges that disorderly motions in the circulation, obstructions, and cohesions "are only secondary *Causes* that accidentally aid and promote some other *Peculiar, Latent* and *Essential Cause*, which I suppose must be joined with them." To learn what this might be, one must turn to "modern Discoveries and Microscopical observations."[68] The "Original and Essential Cause," these observations suggest,

> may possibly be some certain species of *Animalcula* or wonderfully minute living Creatures, that, by their peculiar Shape, or disagreeable Parts, are inimicable to our Nature; but however capable of subsisting in our Juices and Vessells, and which being drove to the Lungs by the Circulation of the Blood, or else generated there from their proper *Ova* or Eggs ... as in a proper *Nidus* or Nest, and being produced into Life, coming to Perfection, or increasing in Bignesse, may by their spontaneous Motion, and injurious Parts, stimulating, and perhaps wounding or gnawing the tender Vessels of the Lungs, cause all the Disorders that have been mentioned. (p. 51)

This hypothesis would seem strange to persons "who have no Idea of any living Creatures besides what are conspicuous to the bare Eye." But those who have studied the "Machinerie" of the universe, who have considered the minute products of divine craftsmanship, and who have "consequently considered the new World of Wonders, that Microscopical Observations have opened to our View," would see its sense. Marten also believed in universal habitation. "There is scarce a single Humour in the Body of Man, or of any other Animal, in which our Glasses do not discover Myriads of Living Creatures." The smallness of these creatures, which may exist even below the threshold of microscopical perception, indicates that their eggs must be even smaller and capable of drifting through the air, settling in our food, and being sucked in with every breath.

> Thus one species of *Animalcula*, by means of their wonderful Smallness and injurious Parts, may instantly offend the Brain and Nerves, and cause Apoplexies and sudden Death, whilst other Species may produce the Plague, Pestilential or Malignant Fevers, Small Pox &c. and others again Chronick

[68] Marten, *New Theory of Consumptions*, p. 50. Subsequent references to this work are cited parenthetically in the text.

Diseases such as Hypochondriack Melancholly, Vapours, Scurvy, Gout Rheumatism, Evil, Leprosy, Consumption, &c. (p. 60)

Animalcula probably swarm in flocks, like gnats and fish. The spread and communication of epidemic diseases is thus "perhaps more explain'd" by this theory than by any other. The specificity of disease is also well accounted for. Marten quotes with approval another animalculist, a Dr. Oliver, who finds that "Diseases keep regular Types, and have particular Attributes that distinguish them one from the other, as the Seeds of Plants do their particular Species" (p. 65).

Like Bradley, who observed that smoke, brine, acids, and some plant juices are lethal to some macroscopic and microscopic insects, Marten turned his attention to the question of therapy. He noted that sulfur power (also recommended by Andry) has only a temporary effect against itch mites, and that when mercury is employed against syphilis (which Andry too observed to be less than wholly effective), it is probable that a few animals or their eggs will escape and breed again, and the symptoms break out with renewed force, a phenomenon we should not ascribe to "a renewed Fermentation of the Venereal Humour" but view as the reproduction of living beings. Remedies applied to the surface may only drive the animalcula inward, where their effects are more deleterious, converting syphilis into grand pox, for example (p. 70).

Marten noted that only in true plagues and pestilences can we suppose that the air, our food, and our water are full of dangerous animalcula and their seeds. In the case of endogenous smallpox, the organisms are probably communicated hereditarily, and, in the case of consumption, the sweat and breath of infected persons will help to disperse the animalcula present in the blood and juices (p. 80). Casual conversation is not usually sufficient to transmit the disease; the few animalcula that are transferred may die, and "some Persons are of such an happy Constitution, that if any of the *Ova* of the inimicable minute Animals that cause a Consumption, happen to get into their Bodies, they may likewise be quickly forc'd out again, through some of the *Emunctories*, before they are produc'd into life; or else be wholly destroyed" (ibid.)

Hooke's account of the water gnat and its development from worm to fly in the *Micrographia* gave Marten, as Hooke's and Swammerdam's accounts had given Andry, a basis for suggesting that, at different stages of their life cycles, the animalcula are capable of different destructive actions. As a worm, the consumption animalcule is perhaps responsible for the early symptoms, such as a slight cough, an elevated temperature, and the obstructions that produce the characteristic tubercules of the disease. Passing to their next stage, the animalcula return to the blood from the lungs; here their gnawing stimulates and wounds the small vessels and leads to high fever, convulsions, and malaise. In a sweat, the animalcula are ejected from the body and temporary relief is obtained until more enter the bloodstream from the diseased lungs. Thus is the intermittence of fevers explained (p. 86).

Natural immunity and resistance are treated only in passing by these authors: the "Royal Experiment" in inducing smallpox immunity through vaccination by Sir Hans Sloane, who succeeded Newton as president of the Royal Society, was performed only in 1721. Marten paints a good picture of the coexistence of man and microorganism. Our blood and juices may always contain great quantities of animalcula. But our general good health—soundness and tone—may prevent them from doing any harm, and so long as our normal secretions and evacuations take place, the animalcula do not remain long enough to do any harm or to achieve the stage of maturity at which they are particularly vicious. When the animalcula irritate our nerves, they may be shaken off and eventually expelled. Only when they find a suitable nidus and are able to make themselves at home do they do us any true injury. By 1730, an attempt to explain acquired immunity in animalculist terms had been made, however. In his account of eruptive fevers, Thomas Fuller attempted to consolidate the theory of generation with the theory of endemic disease, arguing that ova of various kinds "productive of all the contagious, venomous Fevers we can possibly have as long as we live" are implanted in our substance at the earliest stage of life and that these, when fertilized by the male, active cause, bring forth disease—their own "morbid Foetus." The depletion of seeds explains why those who have had a disease such as smallpox or measles once cannot contract it again.[69]

At the end of his treatise, Marten observes that his animalculist account is an alternative to mechanical theories of disease, which have behind them the authority of Newton. Admitting that his account is speculative and difficult to prove directly, he says that he is nevertheless puzzled

> that the Learned Gentlemen of our Profession, who have so excellently well acquitted themselves in mechanically accounting for many Distempers, upon the grand Philosophick Principle of *Sir Isaac Newton*, *viz.* that of Attraction or Gravitation, or the Universal Tendency that one part of Matter has towards another, have not at the same time consider'd what Injuries the Body of Man may receive from the spontaneous Motion of voluntary Agents or *Animalcula* in our Fluids & small Vessels, which *Animalcula* can hardly be

[69] Thomas Fuller, *Exanthematologia* (1722), in Silverstein, *History of Immunology*, p. 14.

supposed to regulate their Motions by Rule and Compass, but act and move according to the natural Instinct, the Divine Author of all Beings has implanted in them.[70]

Andry had made a similar complaint earlier about the fashion for chemical theories:

> Acids and Alkali's are put to too many uses, and the Daily abuse of that Doctrine by the half-way Learned, is a thing to be lamented. Tis an Induction ill drawn from some Experiments of Chymistry, which they unite with Descartes's Philosophy: They borrow the Corpusculums and the connexion of Matter from this Philosophy, to which they join the Acids and Alkali's, which Chymistry discloses to 'em and believe that by this means they have found the Key and Secrets of all Physick.[71]

In Daniel Defoe's Journal of the Plague Year, which pretends to be a first-person account of the plague of 1665 but was actually written and published in 1722, there are echoes of the controversy between Bradley and the Newtonian Richard Mead, author of a treatise on gravitational diseases caused by the sun and moon. The narrator accepts the idea of an infectious effluvium—"some certain steams or fumes" contained in the breath, sweat, or "Stench of the Sores" of the sick person—as though the alternative were simply the direct agency of God, "an immediate Stroke from Heaven, without the Agency of Means." But he cannot accept that the infection is carried "by the Air only," which spreads the disease "by carrying with it vast Numbers of Insects, and invisible Creatures, who enter into the Body with the Breath, or even at the Pores with the Air, and there generate, or emit most acute Poisons, or poisonous Ovae, or Eggs, which mingle themselves with the Blood, and so affect the Body." The air must rather contain a poisonous emission; the other hypothesis is, he says, "a Discourse full of learned Simplicity, and manifested to be so by universal Experience."[72]

One cannot speak of a generally accepted account of contagion in late-seventeenth-century or early-eighteenth-century protoscience any more than one can speak of a generally accepted account of contagion: one can only note which theories attracted particular attention and which were held up to particular ridicule. Like spermaticist preformation, animalculist theories came in for their share of both. This had little

---

[70] Marten, New Theory of Consumptions, pp. 89–90.

[71] Andry, Breeding of Worms, pp. 104–5.

[72] Defoe, Journal of the Plague Year, p. 75. Defoe is here transcribing Mead, who says that "some authors have imagined Infection to be performed by Means of Insects, the Eggs of which may be conveyed from Place to Place, and make the Disease when they come to be hatched," but finds this view to be (for reasons he does not make clear) unsupported by observation. Cf. Mead, Short Discourse.

to do with their being overly "speculative." Every available account was equally speculative, and the prestige of Newtonian medicine of the sort defended by Mead was not negatively affected by its being hypothetical. The Bradley-Marten view even had considerably better reasoning behind it than gravitational illness. But animate contagion was a thoroughly unsettling idea. Few persons writing on the subject still believed that plague was a direct punishment by God, and not the result of a secondary cause. For Newtonian physicians, and for mechanists in general, it could be regarded as an unavoidable by-product of the mechanical order. But if living creatures rather than small particles were the cause, then, with spontaneous generation rejected, plague was doubly intentional: its symptoms were the results of intentional agents—the animalcula—and their existence represented the preference of God, who had surveyed the original Creation, including the microscopical plague animalcula, and found it good. All this hardly squared with the story of the Fall, dominion over the earth and its creatures, or providentialism. Hartsoeker, who was not orthodox, put a gloomy functionalist interpretation on everything: "In fine, it would seem that all Animals were made to serve for Food to one another; the great ones eat the little ones, and are eat up by them."[73] But such views could perhaps only be betrayed by philosophers, not expressed as positive doctrine.

Beyond its general aura of paradox, the animalculist account of disease was associated with charlatanry. It had, after all, an empirical dimension to it, unlike competing theories, and human mechanical ingenuity was able to exploit this dimension effectively. As John Freind described it in his history of medicine, some charlatans "give out that they can draw snakes and lizards out of their patients' noses, which they seem to perform by putting up a pointed iron probe with which they wound the nostril, 'till the blood comes: then they draw out the little artificial animal composed of liver, &c. . . . Others pretend to get out worms, which grow in the ear or roots of the teeth. Others can extract frogs from the under part of the tongue."[74] Defoe had reported in his *Journal of the Plague Year* that he had heard it said that plague "might be distinguished by the Party's breathing upon a piece of Glass, where the Breath condensing, there might living Creatures be seen by a Microscope of strange, monstrous and frightful Shapes, such as Dragons, Snakes, Serpents and Devils, horrible to behold."[75] Andry had as much as said that this was possible. But Defoe remarks—reminding us that he

[73] Hartsoeker, letter to Andry, 26 February 1699, in Andry, *Breeding of Worms*, p. 215.

[74] Freind, *History of Physick* 2:65–66.

[75] Defoe, *Journal of the Plague Year*, p. 203.

was not writing a contemporary account—that "this I very much question the Truth of, and we had no Microscopes at the time, as I remember, to make the Experiment with."

The exploitation of the idea by medical frauds with microscopes was most notorious in the case of M.(onsieur) A.C.D. As the story was recounted by Jean Astruc in his treatise on venereal diseases, this man, whose name was really Boile, stated in his *Système d'un medecin anglois* (1726), that all diseases were caused by distinct species of minute animals in the blood, each of which had a distinct species of antagonist in the form of another kind of minute animal. He represented himself as able to identify the cause of the disease and to prescribe the appropriate therapeutic agent. According to Astruc, he built a microscope in a kind of zigzag form with many mirrors, which he claimed to have constructed according to the principles of Newton's catadioptric telescope.[76] At the end of the farthest tube, M.A.C.D. put a few drops of blood from the sick person. Adjusting the branches of the microscope, he demonstrated a quantity of little animals swimming in a liquid; adjusting the instrument again, he explained that he was adding drops of another fluid containing their natural predators, and, after a few minutes, he revealed the disappearance or death of the first sort. This procedure, which was based on the quick substitution of tubes of liquid containing pond water or other infusions, was extremely convincing, even, as Astruc says, in a city as medically sophisticated and knowledgeable as Paris. M.A.C.D. grew rich from his cures but was eventually exposed and fled. In 1727, he published a sequel to the system, in which he claimed that he had received letters from over thirty persons of distinction who had been thoroughly convinced of the System of Insects by the first essay.[77] He offered to return to Paris and to cure the poor in order to remove the charge of venality and prove his sincerity, but he did not.[78] Astruc draws the moral: "Thus the fables with which some had let themselves be infatuated were exposed, and Medicine happily avenged was reestablished on her old laws."[79] A description of the exact structure of the trick microscope was published by Vallisnieri, the source being Bernard de Fontenelle or someone in his circle.[80]

M.A.C.D. had pretended to be following the system of an unnamed English doctor in proposing invisible insect causes for every disease, not only the venereal diseases, poxes, and fevers, but headaches, impotence, and insanity. He expressed the modern microscopists' view that "all

---

[76] Jean Astruc, *De morbis venereis* (1731), 2:2, in Belloni, "Charlatans," p. 583.
[77] M.A.C.D., *Système d'un medecin anglois.*
[78] Ibid., pp. 7–9.
[79] Belloni, "Charlatans," p. 584.
[80] Ibid., p. 585.

nature is animated" and denied the possibility of generation from corruption: everything came from seeds.[81] Mineral medicines and herbal
simples, he said, were efficacious only insofar as they are actually full of
minute animals and their eggs.[82] The usual infusion experiments, which
showed the development of large numbers of animalcula in standing
liquids within a few days, proved this admirably.

Here, then, a certain philosophical coherence was evidently achieved
with some effort; like many scientific frauds, Boile may have been convinced that his system was essentially correct, though he lacked any
direct means to prove it. On the other hand, one might have thought
the book a satire or spoof, with its slapdash pictures of causative organisms and excogitated nursery names—supposedly to disguise their true
identity—of the specifics supposed to destroy them: Nicota, Vengarsi,
Houplefudree, Sif, Zerzaille, Van, Are, Tilutte, Blino-Blacmel.[83]

The chief intellectual virtue of the theory of animate contagion was that
it provided an explanation for the self-multiplying capacity of contagious illness. A plague victim turns others into images of the plague like
himself. The early-seventeenth-century accounts of plague expressed
this by conceiving the plague as a person who imprinted the image of
herself, by referring to its scattered "seeds," or by reference, as in Helmont, to an "idea" or "form" transmitted by occult means including
fantasy and fear. The corpuscular theory of Boyle and Boghurst was
least satisfactory in respect of its ability to say how the disease was
spread. To explain why plague might first appear in one section of London and then in other quarters, spreading outward from the city, it had
to propose two essentially unrelated modes of infection—one from noxious particles, one from infected persons—or else take refuge in the
extraordinary subtlety of effluvia and make the originally emitted particles capable of drifting for hundreds of miles. Or, as Boyle suggested,
some multiplicative faculty might be ascribed to the toxic corpuscles
without further explanation. On the animalculist theory, the distribution of plague was explained by analogy with the invasion of an agricultural crop by insects over a given area.

If the specificity of the target and the pattern of transmission were
especially well explained by animalculist theories of infection, a number
of other features made them difficult to take seriously. The trustworthiness of the particular observations on which the theory of animate contagion rested was open to reasonable challenge: Kircher, who allegedly

---

[81] M.A.C.D., *Système d'un medecin anglois*, p. 12.
[82] Ibid., p. 13.
[83] Ibid., pp. 14ff.

played a large role in introducing the theory to England, was no one's idea of a reliable witness. Like Paullini, some writers saw with their microscopes "worms" everywhere, in blood, milk, in all tissues of the human body.[84] If insects were the cause of disease, it was not clear why every human being did not get every disease; everybody is subject to fleas and lice. In the absence of a convincing theory of differential resistance, their very ubiquity tended to exclude them as causes of disease. And even supposing what is doubtful, that more than a few microscopes of the period were good enough to detect, not just infusoria, but bacteria,[85] the fact that bacteria—though motile—do not look like insects, certainly not like tiny scorpions, worms, or locusts, or miniature dragons, and do not exhibit obvious feeding behavior would have left their role undisclosed. There is nothing distinctive in the appearance of a bacterium that would have made it identifiable, among all the other newly discovered life forms, as a dangerous agent. The logic of the insect-poison relationship was also unclear. Poisoning causes rashes, headaches, vomiting, pain, death, and other manifestations of sickness, which insects ordinarily do not—unless they are poisonous ones. Thus John Wilkins in his dictionary categorizes "contagion" and "poison" together as "extrinsical" causes of disease, which "either spreading their efficacy by insensible *Effluvia*: or such as being taken in a *small quantity*, prove *destructive* to life."[86] The "gnawing" or "wounding" attributed to insects and the similarity between certain plant blights and human lesions did not explain fever and suppuration.

And how would these insects get into some people and not others? Leeuwenhoek, who wondered about how animalcula might get into the plaque between our teeth, speculated that this might come from the practice of rinsing beer bottles in ponds. Claims about the ubiquity of invisible insects in food, air, and water were nevertheless still difficult to comprehend. If earlier writers saw animalcula everywhere, later ones, armed with somewhat better microscopes, disproved this ubiquity. Henry Baker observed in 1743 that "Many People have imagined, that *living Creatures* might also be found in the other *animal Juices*: but, after the strictest and most careful Examination, it appears certain that nothing with the least Token of Life is to be discovered by the best Glasses, whether in the *Blood, Spittle, Urine, Gall, Chyle*, or in any of the *Humours*, except the *Semen* only."[87]

[84] Belloni argues in "Micrografia illusoria e 'animalcula'" that all these "worms" were optical illusions produced by the lenses of early microscopes.

[85] Leeuwenhoek, whose claim was the best, may have seen the larger-mouth bacteria and perhaps a few cocci or bacilli.

[86] Wilkins, *Essay towards a Real Character*, part 2, chap. 8, p. 220.

[87] Baker, *Microscope Made Easy*, pp. 166–67.

Nor did the general conception of the economy of nature have any place for these destructive organisms. Those who believed in spontaneous generation saw putrefaction and sores as capable of producing life of low forms, and could not comprehend how life of low forms could produce fermentation, fevers, and putrefying sores. Those who believed in an original divine creation could not comprehend how lethal creatures with whom it was impossible to coexist—as we coexist with predators, parasites, and vermin—should have been created, or why, if these creatures were permanent fixtures of the creation, they should display themselves only cyclically. The competitive, opportunistic view of nature was not an appealing doctrine. As the reality of universal parasitism sank in, physico-theologians from John Ray to Friedrich Christian Lesser busied themselves with explaining and justifying it. Some conceptual resources were available for this task. The idea that good health depends on the purging and leaching of unhealthy fluids supports a notion of benign parasitism; had not Aristotle taught that people with lice get fewer headaches?[88] In 1647, Monconys took note of a controversy over the relation between the presence of small wasps and the ripening of figs, and called attention to the process of caprification described originally by Theophrastus. Either the insects contribute some kind of vital force to the plants, or they play a role in the sexual generation of the fruit, or they draw off superfluous wastes.[89] And Monconys is drawn to the view that plague is a case of benign parasitism gone wrong:

> Nature has provided all animals and plants with an infinity of minute invisible insects, which are for the purpose of sucking and drawing out the corruption and impurities of living things: these are like the emunctories of Nature; and if, by some disturbance, these little animate atoms begin to grow and multiply to an extraordinary degree, there occur Epidemic illnesses which result in the deaths of animals. There have, indeed, been observed a prodigious number of these insects in the buboes of the plague-stricken, which, taking wing, carry the infection far and wide.[90]

The discovery of the true extent of parasitism and hyperparasitism in the animal and vegetable kingdoms lent, then, a concreteness to the vague idea of animate corpuscles. But it was only some proof that organisms could not arise from anything except similar parents that could clinch the argument that living organisms were the cause rather than, as seemed more likely, the products of disease.

The quarrel between "contagionists" and "miasmatists" in the late

---

[88] Aristotle, *History of Animals* 557a.
[89] Monconys, "February 1647," in *Journal des voyages* 1:177.
[90] Ibid., p. 178.

eighteenth century over whether the correct explanation of plague, influenza, childbed fever, and other contagious diseases was malignant air and foul water—in which case local, particular precautionary measures, such as sweeping, washing, and outdoor exercise, were ineffective so long as they could not be purified—or an agent transmitted by direct contact, did nothing to contribute to clarity on the question of microorganisms as causes of disease, a hypothesis compatible with both accounts. The distinction was itself misleading. Bad water contains illness-producing microorganisms like the cholera bacterium, bad food and bad air destroy the body's resistance, and close quarters facilitate transmission, especially of diseases like consumption. From the effectiveness or ineffectiveness of general sanitation measures, little could be concluded about the nature of infectious illnesses.

A germ theory of animate contagion can thus be said to have emerged between 1650 and 1720, supported however only by circumstantial and not by direct evidence. Pasteur's conception of a disease process is not essentially different from that of Bradley or Marten. And, in the form in which they present it, the theory is easily distinguished from earlier accounts of a *contagium vivum*, which do not conceive of the agent of destruction and dissemination as a particular living organism.

Contrary to the usual picture of a world of "dead" or "inert" corpuscles in collision managed by God, pananimism and the sufficiency of natural causes were widespread doctrines in the late seventeenth century. The new prestige of mechanics and chemistry in the Newtonian era was influential in medicine but did not affect these basic convictions. Acceptance of a theory of animate contagion was nevertheless an impossible achievement for a science that had only recently left behind it types and resemblances, *archei* and homunculi, but that did not yet recognize a clear distinction between the living and the nonliving. When Jakob Henle in 1840 noted the stubborn persistence in the face of resistance exhibited by the germ theory of disease up to his time, "since the manifestations of the course of contagious diseases must at all times in fact have led to it," he insisted that the minute organisms to which he ascribed the causation of infectious diseases, whether these were animals or plants, were not themselves seeds *of* the disease but rather causes of the manifestations of the disease—thereby, he thought, distinguishing himself from earlier representatives of the *contagium vivum*.[91] The misconception he identified thereby was certainly one that had been held by such writers on disease as Jean Fernel and probably Fracastoro. And indeed, a long treatise by the powerful Leipzig medical professor

---

[91] Henle, *Treatise on General Pathology*, p. 90.

Christian Lange, *Pathologia animata*, printed in his *Opera omnia* of 1688, which was later considered the definitive statement of the doctrine, did not clearly distinguish between an animistic theory of living contagion and an animalculist one. There are two distinct strata in *Pathologia animata*, whose doctrines were disseminated in the German medical literature by the many students Lange taught and supervised. On one level, Lange argued, following Helmont, that disease symptoms were reactions of the host attempting to expel an invader; on another level, he accepted Hauptmann's suggestion that infestation by hostile living agents might be the cause of the damage. Henle himself failed, it might be noted, to recognize the truth buried in the Helmontian conception of disease: fever and other morbid symptoms such as suppuration are in fact the signs of a counterattack on the hostile organism. But he observed that many earlier investigators had identified the causes of disease naively as "insects," suggesting thereby that it was a taxonomic awkwardness as much as a false understanding of a process that kept the theory of animate contagion just outside the realm of the incredible.

In any event, as Henle pointed out, animate contagion is easily suggested but difficult to prove from positive observation alone. Microorganisms can always be claimed to be parasitic accompaniments of sickness, and even to aid in the diagnosis, just as the spermatozoa could be regarded as parasitic accompaniments and signs of male fertility by convinced ovists. Only isolation and direct experience of the exciting role of the pathological organism could exclude this interpretation. Even so, the idea of specific anti-insect therapies as a means to health had achieved an airing, and the confusing logic of preservative and destructive pharmacological substances was beginning to receive attention. Leeuwenhoek, who, like Andry, was convinced that tooth decay resulted from the destructive action of animalcula in the human mouth, noted that they are killed by hot coffee and that they all die in vinegar, "whence it seems reasonable to conclude, that washing the Teeth and Gums with Vinegar may be a means of preserving them from these minute creatures."[92]

---

[92] Baker, *Microscope Made Easy*, p. 158.

6

The Philosophers and the Microscope

THE MAJOR philosophers of the late seventeenth and early eighteenth centuries were well attuned to the experimental sciences as well as to cosmology and rational mechanics, and they show a range of reactions to the discovery of the microworld. Some, like Leibniz and Malebranche, whose reception was wholeheartedly positive, found in the revelations of the microscope a way of reknitting the unraveling relationships between the natural and the supernatural, situating the knowledge of nature within theological space. The apologetic function and reactive character of much seventeenth-century metaphysics is an inescapable fact for the historian.[1] In its effort to come to terms with what existed, rationalist metaphysics found itself on an awkward footing. But the notion that biology assumed an objective standpoint only with Darwin is incorrect. The physico-theology that Kant attacked as an anthropocentric illusion was itself a kind of paradigm of objectivity; the eighteenth-century notion of the "spectacle of nature," with its connotations of distanced viewing, suggests both something staged by God for our elucidation and a show that proceeds on its own terms.

Physico-theology, the project of reading in the features of the world the existence, presence, and characteristics of a supernatural being, did not begin in the seventeenth century, but it assumed increasing momentum and even developed into a standard format for the presentation of natural history in the first quarter of the eighteenth. While there is no doubt that its texts are sincere expressions of the desire to harmonize religious belief with those aspects of the new science that threatened it, their proliferation had a much broader context. The surveillance and control of atheism, which had once been a matter for individual priests, pastors, and confessors, became, in the age of print, a task for the literati. Philosophical rationalism and skepticism sometimes bolstered but also challenged the Christian edifice, which was built on an acceptance of the miraculous, and Christian philosophers initially attempted to meet such challenges by argument and demonstration rather than by direct appeals to scriptural and priestly revelation or by advancing to a daring, last-ditch fideism. For this saving role, the argument from design

[1] See Loeb, *From Descartes to Hume*, esp. pp. 327ff.

was ideally suited as, in appearance, a revelation without mediation through texts and authorities.

Meanwhile, the enterprise of scientific virtuosity required a kind of theological cover to disarm the charge that investigation of natural causes was by its very nature hostile to religion because it excluded creative and interventionistic activity on God's part and diverted the moral and spiritual quest for perfection into a material and intellectual one. In the background lurked the threat of libertinism, which insinuated itself though a series of easy deductions: whoever studies nature sets a value on nature; whoever sets a value on nature seeks to live according to nature; whoever seeks to live according to nature lives outside the laws of God and men.

The *Religio medici* of Sir Thomas Browne, intended to exonerate the vulnerable class of doctors from the charge of atheism, and the physico-theologies of John Ray, William Derham, Bernard Nieuwenwyt, and their wide band of imitators disseminated detailed knowledge of natural history. At the same time they sought to block the inference to which the study of natural history and medicine seemed inevitably to point: that nature was either radically other than God, neither created nor guided, or else it was identical with God. But as Wolfgang Philipp reminds us, physico-theology was not simply the capture and domestication of science by theological authority, nor a heretical betrayal of the book of Revelation in favor of the book of nature, but both at once.[2] It was a movement of restoration, by means of which nature, a word that conjured up in its semantic field paganism, idolatry, and hedonism, could be portrayed as dependent and passive, and as part of a strictly controlled and regulated system, as a theocratic civil society of a sort, rather than as an alternative to one. It was, at the same time, a large concession to secular interests.

Its books found a foothold and propagated their kind not only because religious faith was still strong, but because positive religion was already on the defensive. The scriptural Word did not, in a world of alternative words, books, and texts, have the weight that it had once had, so that deductions of the facts of immortality and the existence of a unique benevolent God, which were not self-evident, needed either to be drawn from systems of pure reason—witness Descartes's attempt to vindicate the doctrine of survival by proving the soul, through a lengthy and roundabout series of deductions, to be immaterial[3]—or a posteriori, from observation. Often quoted in, or at the front of, natural

---

[2] Philipp, *Das Werden der Aufklärung*, p. 73.

[3] Note that the immateriality of the soul was not a necessary part of Christian doctrine; Gassendi, for instance, argues for a flamelike, material soul.

histories was the statement from Romans 1:20 that the invisible things of God are shown in the things that are made. This anchoring of the study of nature in a citation was perhaps intended less to prove that the study of visible things had a scriptural warrant than to indicate that the Scriptures still had something to say to the moderns. But there was a certain circularity in the method of physico-theology. To prove the existence of God from an inspection of the world, one had to prove that it was a well-ordered mechanical system. Yet the best argument that it was in fact a well-ordered mechanical system was that it was a divine creation. The rationalist texts examined in this chapter do not so much produce an argument as urge a certain image on the reader. From the conflicting, enigmatic, and largely unexpected data of empiricism, they attempt to compose a picture in which God as artificer supports a macro-mechanical world system composed of micromechanisms, one in which determinism, the uniform and ubiquitous operation of laws of nature, supports claims about the uniqueness, power, and benevolence of God.

When Descartes wished to persuade people that the body really is an ensemble of integrated micromechanical devices, he appealed to their preexisting intuition of God's powers; Malebranche preferred to argue that God's powers can be appreciated in the feats of internal engineering of the body. Something of the modern Christian virtuoso had already appeared in the Paracelsian idea of a religiously motivated and empowered observer. Imperfect concealment is a trademark of God, and the success of the physician at playing hide-and-seek with pharmacological substances gives him a fair idea of the game and its rules. In the physician's seventeenth-century counterpart, this virtuosic machinery was similarly busy; he was constantly reading up from observation to theology, and down from theology to observation, and generating a considerable literature in the process.

The details of the charge against the naturalist and its answer are admirably set out in Robert Boyle's essay "The Christian Virtuoso." Boyle's argument is that science is only contingently associated with libertinism. He acknowledges the "profane discourses and licentious lives of some virtuosi, that boast much of the principles of the new philosophy."[4] And he admits that science may be perverted and misdirected by a libertine: "[I]f the knowledge of nature falls into the hands of a resolved atheist or a sensual libertine, he may misemploy it to oppugn the grounds, or discredit the practice of, religion."[5] But, says Boyle, there is an intrinsic and normal connection between virtuosity and Christian faith. An increase in knowledge can only rule out more

[4] Boyle, *Christian Virtuoso*, in *Works* (1772 ed.), 5:514.
[5] Ibid.

and more definitively the hypothesis that the world is produced by "so incompetent and pitiful a cause as blind chance or the tumultuous jostlings of atomical portions of senseless matter."[6]

In Boyle's physico-theology there is a sense of unease, a deep worry about the incompatibility of science and religion, that needs to be laid to rest. But what is the connection—the encouragement to disorderly living aside—between the exact knowledge of nature and unbelief? It is not wise, as Boyle was well aware, to spell this connection out too explicitly in an apologetic treatise. The careful laying out of heresy or heterodoxy under the pretext of trying to refute it is the old strategy of atheists who, having shown people how to think, then go through the empty motions of refuting themselves. Boyle could not face his subject too squarely lest he make a gift of arguments to his opponents. Nevertheless, various connections are postulated. The pagans were interested in nature, and made it the basis of their moral philosophy. The investigation of nature is a distraction and even a kind of idolatry. The discovery of invariable laws that relate causes to effects, or prior states to posterior ones, encourages generalization to a belief in an exceptionless, miracle-free course of nature. The familiarity with natural wonders and marvels lead the person of wide experience to suppose that biblical wonders and marvels are part of the ordinary course of nature and to deny their special role in establishing the authority of the Church. And, finally, physicians who deal with the dying and the dead and who perform autopsies do not appreciate that death is a pathway to a better world, that its moment is chosen by God, but see only something that reminds them, as it did Descartes, of a broken clock.

Boyle needed to subvert these patterns of inference by reinterpreting scientific experience as a new form of worship. The virtuoso who "searches deep into the nature of things" and who observes "exquisite contrivances ... admirable co-ordinations, and subordinations" that "lye hid from those beholders, that are not both attentive and skilful"[7] is truly able to appreciate the creative and regulatory powers of God. The old language of signatures and traces persists, only it is now fixed to a notion of interior, small-scale differentiation: God has stamped

> upon divers of the more obvious parts of his workmanship such conspicuous impression of his attributes. . . . For . . . besides the impresses of his wisdom and goodness, that are left, as it were, upon their surfaces, there are a great many more curious and excellent tokens and effects of divine artifice, in the hidden and innermost recesses of them; and these are not to be discovered by the perfunctory looks of oscitant or unskilful beholders; but require, as well

[6] Ibid.
[7] Ibid., p. 518.

as deserve, the most attentive and prying inspection of inquisitive and well-instructed considerers.[8]

The search for them was not simply permitted, but mandated, if one were not to miss God.

This talk of the visual world as an empire of riches to be surveyed and known, as well as the disparagement of ordinary sense experience, situates Boyle's treatise with respect to the science of the past. As Hans Blumenberg, who describes Aristotelian science as resting on the "postulate of the visible," remarks, the developing science of the early modern period invoked two rhetorical tropes, spatial and temporal: the widening of the objective horizons of knowledge, and the discrediting of the past and its knowledge as prehistory. But this deliberate effort to destroy the old spatiotemporal framework left modern science somewhat unsure where it was. "In this rhetoric," Blumenberg says, "metaphors like sea voyage and the discovery of unknown lands, the crossing of borders and the breaking through of walls, of microscopic and telescopic optics play a preferred role. They show the problematic of orientation in a reality for which standard measure, scope, and direction were almost entirely lacking."[9]

The problem of retemporalization manifested itself in, for example, Bacon's ambivalent attitude toward ancient observational science and in the need and temptation to draw on earlier authors even while maintaining that it was necessary to do all their work over. The problem of recalibration has been studied up to now mainly in connection with the opening out of geographical and celestial boundaries. The knowledge that the inhabited earth is much larger and much more varied than might have been thought induced a disturbing and exhilarating moral and theological relativism. The discovery that stars exist at different distances from the earth, extending out perhaps infinitely in all directions rather than being pasted on the surface of the highest crystal sphere, has been held responsible for a sort of miserable vertigo in some minds, a delirium of ecstasy in others.[10] Post-Renaissance geographical and astronomical decentering was balanced by nationalistic technological and intellectual centering. But military and mathematical mastery of new territory threatened as well as enhanced confidence. Their new astronomy, produced by their own mathematical intelligence, taught the moderns that they could make a better picture of the world than the ancients had, perhaps as good a picture of their world as God. But works like Thomas Burnet's *Sacred Theory of the Earth*, with its geo-

---

[8] Ibid., p. 516.
[9] Blumenberg, *Die Lesbarkeit der Welt*, p. 68.
[10] See the classic study by Koyré, *From the Closed World to the Infinite Universe*.

theological parallelism, exhibit the compensatory attempt to restore the physical importance of the earth and its status as the theater of events of supreme moral importance. The revisions implied that the recalibration with respect to the large and the broad had some parallels with respect to the small. The texture of nature was finer, its subtlety more extreme, than anyone had thought. This discovery proved the special nature of human beings, who, alone among creatures, were able to perform feats of seeing and understanding that transcended the mundane. At the same time, it raised questions about the extent to which these beings and any others could be considered particular beneficiaries of a providential scheme.

Attention to the microworld helped compensate for the loss of the greater cosmos as an image of divinity and spiritual order. In the new astronomical systems of the sixteenth and early seventeenth century, the critical feature was not the relegation of the earth to planetary status but the breaking through of the enclosing shell of the fixed stars, which reduced to nonsense the orientating notions of up and down, heaven and earth, with their connotations of excellence and baseness. There was now only out—toward other suns and satellites, which might even possess their own ensouled inhabitants—and in, toward this arbitrarily designated earth, which might be out for other intelligent beings in turn. To convert a loss into a gain, it was necessary to stress the discrepancy between divine immensity and terrestrial puniness. In the first generation of philosophers of the minute, a similar strategy was set in motion. A providence directed toward human beings specially and the worship of a being who directs and intervenes gave way to deistic worship of the creator of a system that needs and can sustain no intervention after creation. The microworld was as much a part of this expanded system as the macrocosmos. I noted earlier the tendency—mentioned by Margaret d'Espinasse—of seventeenth-century natural philosophers to favor the microscope and the study of living creatures over the telescope and the theory of the heavens. True to this pattern, Leibniz pointed out that, for knowledge of nature, and for the purposes of religious inspiration, "telescopes are far from being as useful and from revealing the beauty and varieties of nature which microscopes reveal." What the telescope takes away the microscope returns:

> Nothing better corroborates the incomparable wisdom of God than the structure of the works of nature, particularly the structure which appears when we study them closely with a microscope. It is for this reason, as well as because of the great light which could be thrown upon bodies for the use of medicine, food and mechanical ends, that it should be most necessary to push our knowledge further with the aid of microscopes.[11]

[11] Leibniz, "Reflections on the Common Concept of Justice," in *Leibniz*, p. 566.

Another example. The microanatomist and clergyman Nehemiah Grew's *Cosmologia sacra: or, A Discourse of the Universe as it is the Creature and Kingdom of God* (1701) was written, the introductory remarks explain, to refute the "many lewd opinions" of antiscripturalists writing in Latin, Dutch, and English, but especially those of Spinoza. Individual chapters are dedicated to the corporeal world and the uses, or functioning and purposes, of ether, air, seas and rivers, currents and tides, and to corporeal bodies, life, truth, mechanism, and divine providence. Infinity held no terrors for Grew and seemed to occasion rather a feeling that was a precursor of Kant's experience of the natural sublime, in which fear in the face of overwhelming magnitude is transmuted into pleasure caused by respect for the Creation and the conviction of personal security. "As there is no *Maximum* whereunto we can go, but God only, so there is no *Minimum, but a Point; which* hath no Dimensions, but only a Whereness, and is next to Nothing. . . . Principles of bodies may be Infinitely Small, not only beyond all naked or assisted Sense; but beyond all Arithmetical Operation or Conception. . . . Ten thousand Seeds of the Plant called *Hart's-Tongue* hardly make the Bulk of a Pepper Corn," but each has "parenchymous" and ligneous parts, fibers, principles, and atoms, which, "being but moderately multiplied by one another, afford a hundred thousand millions of Formed Atoms in the Space of a Pepper-Corn."[12] God's government is

> Universal. Reaching not only to Celestial but Terrestrial Worlds; and among them, this we live in. Not only unto Things of greater Moment: but unto those also, which seem to us to be the most Casual, and the most Trivial. And so unto every Work, Thought, Motion or Contingent: or the Operation, as well as the Being of every vital Principle, and of every Atom. And therefore not only unto the Proximate Effects of Things; but unto all others the most Remote.[13]

Grew was interested not only in the individuality of the human face and its meaning; he studied with a lens the patterns of the pores and ridges of the human hand as well, and saw in their distinctness and uniqueness proof that God had concerned himself with every particular and every detail.

But where is the dignity in studying insects? The earlier naturalists had looked to these diverse and surprisingly available creatures for instruction through their exemplary traits and habits. Now it was their structure that informed. Earlier the moral lesson they taught had helped establish the worthiness of these otherwise negligible and base objects.

[12] Grew, *Cosmologia sacra*, p. 12.
[13] Ibid., p. 86.

Solomon had instructed the sluggard to go to the ant, and sixteenth-century observers remind us repeatedly how much is to be learned from observing their care for their offspring, their diligent use of their senses, their tireless pursuit of nourishment, their digging and tunneling. The shyer bird and mammal do not often give us this opportunity for direct inspection; the lesson in conduct and deportment comes more easily from the ant and the grasshopper. Yet small-animal valorization of the sort initiated by Pliny, who had said—in a passage quoted obligatorily and so ad nauseam by physico-theologists—that they show the divine wisdom better than great ones, did not ensure the valorization of the time and trouble of studying them, and the justification for natural history as an enterprise remained highly problematic.

Thus it was from a defensive posture that Thomas Moffett, at the end of the sixteenth century, strove to establish his subject as meaningful and attractive. The work on which he was now embarked was, he explained, a completion of a book begun by the natural historian Pennius,[14] whose torn letters were part of an abandoned or wrecked manuscript that Moffett had recovered and pasted together. Three difficulties stood in his way. First, the subject was new and difficult: "I saw that insects are hard to be explained both in respect of the unusualness of the subject and also of the sublime or rather supine negligence of our Ancestors in this point." Second, it was hard to describe insects in a beautiful style. Third, he had to face the criticisms of his friends, who considered him to be wasting his time in the study of imperfect living creatures and to be demeaning himself in taking on someone else's unfinished work. Moffett scornfully remarked on the vulgar excitement over large, new animals, such as crocodiles and elephants, complaining that "no man regards Hand-worms, Worms in Wine, Earwigs, Fleas; because they are obvious to all men, and very small, as if they were but the pastimes of a lascivious and drunken Nature, and that she had been sober only in making those huge and terrible beasts. . . . Farewel then all those that so much esteem of creatures that are very large, I acknowledge God appears in their magnitude, yet I see more of God in the History of Lesser Creatures."[15]

Moffett's *Theater of Insects* appeared later as the fourth volume of Edward Topsell's *History of Four-Footed Beasts and Insects* (1658), a compendium said by the author to be "necessary for all Divines and Students, because the story of every Beast is amplified with Narratives out of *scripture, Fathers, Philosophers, Physicians and Poets*" and

[14] This was Thomas Penny, fl. 1570, a student of Conrad Gesner, who collected material on insects from four hundred authors; see Gunther, *Early Science in Cambridge*, pp. 337–38.

[15] Moffett, *Theater of Insects*, preface.

adorned with "*Hieroglyphs, Emblems, Epigrams*, and other good *Histories*."[16] The quotational, symbolic, and literary approach is still evident in the preface to the Moffett volume by Theodore de Mayherne. The reader is promised delights, gratifications, moral and useful knowledge. He will learn of the bottle of the bees, "which is the receptacle of honey sucked from flowers"; the horny sting, "full of ravening poison"; the trumpet of the gnat for sucking blood; the "small proboscis of Butterflies wreathed alwaies into a spiral line . . . their extended large wings painted by nature's artificial pencil, which with paints cannot be imitated." You will see, he says, the muscles of the horns of the rhinoceros beetle, how the web of the spider is thrust out of her body, the devouring foreteeth of the locust "like chizels," the sharp spears in the mouth of the spider. "What a pleasant spectacle this will be when the artificial hands carefully and curiously guide the most sharp pen-knife and very fine instrument by direction of the sight!"[17]

Besides this philosophical instruction, there would be lessons in monarchy (bees and their king); in democracy (ants); in household economy (spiders and silkworms); in engraving (woodworms); in maternal and paternal care (oil beetles); and in architecture (bees). "They do better in my opinion who observe the Pismire and grow rich by following his manners in labor, industry, rest and study."[18] We should emulate the study habits of ants? The reference point is always the human. What engages the writer's attention throughout the text are the similarities between human and insect forms of social behavior and organization, the tools and equipment used by insects, the nomenclature and uses of insects, and the damage and suffering wrought by insects, preeminently the flea, with its "hollow trunk to torture men, which is a bitter plague to maids, and is the greatest enemy to human rest." Moffett even reminds us of the enmity between insect and book: "[T]hose beasts that are the greatest enemies to the Muses and their darlings, I mean the Moths" with their "greedy bellies" and "iron teeth" that devour the printed page. But the reference, through the human, is to God too: "In the work of these Spiders," Moffett says, "if you consider the wouf, the skeins of yearn, the trendle, the shuttle, the comb, the wool, the distaff, the web, either you will see nothing, or you will see God, insensible, yet really perfecting all these things." The book is about the harmony of the Creation. Insects are recognized as destructive and troublesome, but there is no hypostatization of opposed interests; those accustomed to

---

[16] Topsell, *History of Four-Footed Beasts*, p. 341.
[17] Moffett, *Theater of Insects*, epistle dedicatory.
[18] Ibid., p. 1078.

the rhetoric of pest control are surprised by the mild attitude of acceptance.[19]

But as the descriptive sciences of anatomy and natural history gained momentum, the note of worry grew increasingly strong. The issue was no longer just whether an aficionado like Moffett was wasting his time on frivolous subjects, as his friends thought; the question was raised on the wider social and institutional scale. To what purpose was this observation of nature? And in the discourse of Christian virtuosity, what was perhaps the real motive—*curiositas*, wonder, admiration—was denied, or rather transformed into something else. To Swammerdam, perpetually on the verge of a crisis of confidence about his work and its value, God revealed himself in a butterfly: "the wings [are] bigger than the body, and with all their colors, which are chiefly red and black. About the edges, however, here and there some yellow, sky-blue and white spots are seen most elegantly combined, which exhibit to us, though faintly, and as it were by shadows, the inexhaustible treasure of the Great Creator's treasures."[20] And Malebranche—the philosopher who said that our visual perception and our action occurs only in and through God—rhapsodized in his turn about insect form and color:

> The tiniest gnats are as perfect as the largest of animals. The proportion of their members is as correct as that of other animals, and it even seems as though God has willed to bejewel them in compensation for their lack of size. They have crowns, plumes, and other attire upon their heads against which anything invented by the riches of men must pale; and I can assert with confidence that those who have used only their eyes, have never seen anything so beautiful, so fitting or even so magnificent in the houses of the greatest princes as what can be seen with magnifying glasses on the head of a simple fly.... So much beauty is found concentrated in so small a space; although they are quite common, these animals are nonetheless remarkable.[21]

Though enthralled, Swammerdam was also harrowed by the insect: he wrote to his patron that "the omnipotent finger of God is presented to you in the following sheets, in the anatomy of a vulgar and loathed insect, the louse." Though the louse is of no advantage to the body, "yet it is able to raise our thoughts to God; so that by seriously contemplating the divine Majesty, and the glittering rays of his miracles, in this

---

[19] Contrast the stance of the modern agriculturalist or grocer. A professor of food science writes to the *New York Times*, 6 February 1992, "Food treated with ionizing radiation will . . . look fresher and not harbor parasites, molds, bacteria or insects or insect eggs that might harm us."

[20] Swammerdam, *Book of Nature*, part 2, p. 19.

[21] Malebranche, *Search after Truth*, p. 31.

little animal, we may, with most submissive humility, change and contract our vain pride into as small a point."[22] The automatisms of insect behavior were especially beloved by the Christian naturalists, for they showed them God's thought and action in the pseudointelligence and pseudowill of the creature. Leeuwenhoek decided that the bee was a mindless machine, driven hither and thither by patterns of light falling on its compound eyes, and, in keeping with the general trend of seventeenth-century metaphysics toward theological voluntarism and mechanism, insects were increasingly treated as automata. Bees, like domestic fowl, Swammerdam said, cannot help making their nests and nourishing their young, and he protested against the anthropomorphization of the natural-history writers. There was no authority for the prevalent and common opinion that bees have a society, that their government is "carried on with the scepter of prudence and judgement, under law and with rewards and punishments; for in truth all that order which we so much and not without reason admire, is impelled by nature, and is only designed for the propagation of their species." Bees exist only to make more bees: the hive is not a model of civil society, but a giant nursery.

The facts of Swammerdam's life reveal some of the difficulties of natural history as a career for a seventeenth-century bourgeois, both from the socioeconomic and institutional perspective and from the personal and individual perspective. Swammerdam was born in Amsterdam in 1637, the son of an apothecary and naturalist whose cabinet was stocked with animals, insects, vegetables, and fossils attractive in the usual way to tourists and noblemen. The younger Swammerdam was early on employed in cleaning and arranging the curiosities and soon began to seek information and to add to his father's collection on his own. His biographer Hermann Boerhaave tells us that, as a youth, "He ransacked . . . the air, the land, and the water; fields, meadows, pastures, corn grounds, downs, wastes, sand hills, rivers, ponds, wells, lakes, seas, and their shores and banks; trees, plants, ruins, caves, uninhabited places, and even bog-houses, in search of Eggs, Worms, Nymphs, and Butterflies; in order to make himself acquainted with the nests of insects, their food, manner of living, disorders, changes or mutations, and their several ways or methods of propagation."[23] He knew more about insects, Boerhaave says, than all authors of preceding ages put together. In Leyden he began to study medicine, under Franciscus Sylvius, or François de la Boe, and became acquainted with Nicholas Steno and Jan van Horne. He was at first a friend, but by 1672 an

[22] Swammerdam, *Book of Nature*, part 1, pp. 37–38.
[23] Ibid., p. ii.

enemy, of de Graaf, whom he would unsuccessfully denounce to the Royal Society as a plagiarizer of his own work. In Leyden he worked on the improvement of techniques for preserving anatomical specimens for medical instruction to avoid the necessity of repeated autopsies. In 1664 he discovered valves in the lymphatic vessels, visited the Loire region to study the day-fly, and, while, stopping off in Paris, developed a friendship with Melchisedech Thevenot, the former French king's minister at Genoa. He returned to Amsterdam and studied the spinal marrow, working at a method for injecting vessels in the body with red and yellow wax in order to solidify them and render them stiff for section, an important anatomical technique that requires cleaning the surrounding tissues away with a solvent.[24] He finished his medical degree in 1667 with a dissertation on respiration. To overcome the problem of different layers of tissue becoming confused by wax injection, he blew them apart with tiny tubes and allowed them to dry. He excelled in techniques of preparation presumably derived from the undertaker's and embalmer's arts, including soaking in turpentine, balsam, or mastic, in addition to the usual preservative, spirits of wine.

Returning to insect anatomy, his real love, he discovered in 1668 the preexistence of the butterfly in the caterpillar "by means of instruments of an unconceivable fineness." In 1669 he published in Dutch his *General History of Insects*, which was quickly translated into Latin and often reprinted, followed by treatises on the uterus and, in 1675, on the day-fly. At this point the elder Swammerdam, who had been unenthusiastically financing the researches of the son, began to insist that Jan, who was now thirty, achieve financial independence by taking up the practice of medicine. Swammerdam resisted; his father cut off his allowance. Exhausted and sick, he requested a country vacation where, in the privacy of his rural retreat, and tempted by the abundance of insects, he began to collect again. Thevenot offered him sanctuary in France, which he refused; he had already refused in 1668 on religious grounds the offer of the Catholic grand duke of Tuscany to take a post at the court of Florence and care for his collection—an offer that had been accompanied by the promise of a very generous sum for the collection. He next tried unsuccessfully to placate his father by setting the latter's cabinet in order, meanwhile beginning anatomical investigations on the womb, ovary, and spermatic vessels, and quarreling with de Graaf. He studied the pancreas, liver, and melt of fish. In 1673, Antoinette Bourignon, the charismatic Protestant quietist and mystic, entered his life with her message of the vanity of worldly achievement.

---

[24] Vessels were also injected with mercury or resin; see Bracegirdle, *History of Microtechnique*, p. 11.

After entering into correspondence with her, Swammerdam continued his work on the bee but in a state of growing distress. According to Boerhaave, he "bent his endeavors more particularly to suppress the unruly passions of the mind, and above all the insatiable ambition which makes us so desirous of a superiority over others, and which therefore, as the root of all evil, he was utterly desirous to extirpate and destroy."[25] Yet he worked tirelessly, from six to twelve every morning, ceasing only when his eyes were wearied by the bright summer light and his microscopes. He ingeniously admits, Boerhaave tells us,

> that his treatise of Bees was formed amidst a thousand torments and agonies of heart and mind, and self-reproaches, natural to a mind full of devotion and piety. On one hand his genius urged him to examine the miracles of the great Creator in his natural productions, whilst on the other, the love of that same all-perfect Being deeply rooted in his heart, struggled hard to persuade him that God alone, and not his creatures, was worthy of his researches, love and attention.[26]

The bee treatise that forms a part of the *Book of Nature* is admirable both for the care and accuracy of Swammerdam's representations of the proboscis, sting, and eye, and for its definitive clarification of the nature of the hive and its personnel, which consists of one female at the head and a staff of male drones and neuter workers. Fascinated, like every microscopist, with the insect eye, he decided that with their compound eyes, bees could not really see as we do, for no image, he thought, was formed on the retina. Yet the importance of the visual sense in insects was indisputable: he was surprised to see "how tame and gentle the Fly becomes when its eyes are . . . covered with paint; it suffers itself to be caught every moment, and when it runs or flies, you will see it stumbling everywhere, and when this happens it is driven back, like a ball, by whatever opposes it."[27] He made another celebrated series of observations on the snail and the formation of its shell, which, he said, resembles the formation of bones and teeth in vertebrates and occurs through the secretion of fibers that are hardened by the air from its mucus glands.[28] He saw the red corpuscles in both frog and human blood—an indication of how good his microscopes were.

It was not long after meeting Bourignon that Swammerdam lost interest for good in observation, publication, and worldly accomplishment. He destroyed his treatise on the silkworm; in 1675, Steno sent the illustrations, which Swammerdam had spared, to Malpighi. He resolved

---

[25] Swammerdam, *Book of Nature*, part 1, p. vii.
[26] Ibid., p. ix.
[27] Ibid., p. 216.
[28] Ibid., pp. 69–70.

to sell his collection of insects and anatomical specimens, which now comprised more than three thousand items, to raise enough money to live. Steno tried to arrange another position for him in Catholic Florence, which Swammerdam, after consultation with Bourignon, rejected as venal. He returned home, where the household was breaking up; his father had died still opposed to Swammerdam's manner of life, and there was trouble about the will. Swammerdam fell sick and remained indoors in bed, refusing medicine and fresh air. He wrote to Thevenot, who no longer had anything to offer. He grew sicker, swelling, losing weight, in pain, and finally made his will in January 1680, leaving everything to Thevenot. He died in February. The *Ephemeri vitae*, his last work, a treatise on the day-fly, and the *General History of Insects* were sold by Thevenot's heirs, later falling into the possession of Boerhaave, who reassembled them and issued them in 1737–38 in a bilingual Dutch/Latin edition under the title *Bybel der natuure* (the Bible, not the book), with accompanying microanatomical engravings. The day-fly study is a religious lesson in transitoriness and quietism. The creature lives for four to five hours. Innocent and simple, most of its larvae are gobbled up at birth by fish and devoured by birds swooping at them from the air as they emerge from their cocoons. The day-fly is devoid of reason in order that "He, from whom springs all reason and knowledge, might take upon himself the care of nourishing its prodigy."[29]

Swammerdam's work was accomplished both because of and in spite of his individual misfortunes: his emotional lability, his rigid father, his ineptitude for social and intellectual relations, the lack of recognition from the Royal Society, which ignored him in favor of the gracious, stylish Malpighi, and what Leibniz described as Swammerdam's oversusceptibility to spiritualistic impressions. Boerhaave speaks of the "exactness of melancholia," of an "ardent imagination of passionate sadness which bore him towards the sublime." The passion for collecting and dissecting filled him with guilt, like an irresistible bad habit, and set Swammerdam apart from the other classical microscopists: Hooke, the jack-of-all-trades, Malpighi, with his art collection and domestic life, the stubborn but on the whole jovial Leeuwenhoek, with his shopkeeping, and Grew, the confident clergyman. To the list of provocations Jules Michelet added the insect: the abyss of life appeared in all its depth with its millions upon millions of unknown beings and bizarre forms of organization that no one had dared dream of.[30] The anatomy of the snail was "a labyrinth of miracles." Swammerdam's exaggerated

[29] Ibid., pp. 117–18.
[30] Michelet, "Swammerdam," in "L'insecte," p. 341.

sensibilities found beauty and horror everywhere; nature seemed to present both faces—*naturans and naturata*—to him simultaneously; it was the Creator himself and it was filth.

> How then can we avoid crying out, O God of miracles! how wonderful are all thy works! how beautiful are the ornaments! how well adapted the powers which thou hast so profusely bestowed upon thy creatures! They are all, notwithstanding, subject to decay and destruction; and, with all their perfections, scarce deserve to be considered as shadows of the Divine Nature. It is therefore, with the highest reason, that a certain writer has said, That all nature is over-run, and covered with a kind of leprosy. This is her old garment, which she is one day to throw off, and its heaviness alone is sufficient to weigh down our senses and disturb our reason, in spite of all its efforts.[31]

Swammerdam's strength, says Cole, "lay in observation and experiment. In reflection he was dangerous."[32] His physico-theology was too erratic to give him mental equilibrium, or else his equilibrium was too defective for him consistently to admire rather than being appalled. But for philosophers less troubled to begin with, or with a natural bent toward optimism, his reflections as well as his experiments supplied material. They registered particularly with Leibniz, who appeared to find the accommodation of religion and science easy—perhaps too easy: Leibniz clucked at Swammerdam, Steno, and Pascal, three devout and brilliant minds mostly lost to science, who, if they had only known how to combine the true metaphysics with physics and mathematics, would not have been overwhelmed by guilt and qualms.

The theme of disproportion, or the impossibility of any adequate calibration with respect to both the great and the small, was eloquently worked by Pascal, who used it to undermine rationalist confidence. If science is false, human intellectual endeavor is empty, whereas if it is true, vanity is still humiliated. The world is a point in the cosmos, and "no ideas can come close to grasping it." But a human being is as monstrous and bumbling as he or she is tiny and insignificant; Pascal turns the human-as-measure inside out. "[A] flea offers him in the smallness of its body incomparably small parts: legs with joints, veins in its legs, blood in its veins, humours in the blood, corpuscles in the humours, vapours in the corpuscles, which dividing again into smaller things, he exhausts his forces in these conceptions. . . . I wish to make him see there a new abyss. I wish to paint for him not only the visible universe, but the immensity which one may conceive in nature."[33] The broaden-

---

[31] Swammerdam, *Book of Nature*, part 2, p. 20.
[32] Cole, "Jan Swammerdam," p. 219.
[33] Pascal, *Pensées*, no. 15, in *Oeuvres complètes*, p. 526. The abyss image, like the

ing of perspective left an unsettled feeling: why should God have created so many different species, each with its own niche and purpose? The language of plenitude and harmony was designed to fill this gap, but it did not impress everyone. Not Bayle, for example, who described in his *Dictionnaire historique et critique* the horrors of omnipresent life and the constant destruction of ensouled beings who may well enjoy mentality and intentionality. "The microscope has revealed them by the thousands in a drop of liquid and we would discover more if we had better microscopes. One should not say that these insects are machines—one should rather explain the actions of dogs that way than the actions of ants and bees. There may be more intelligence in invisible animals than in the larger ones."[34]

Meanwhile, the philosophies of Malebranche and Leibniz reflect the attempt to subsume and domesticate the notions of infinity and immensity, to make them part of a system, to be able to say what the divine view of things is even while admitting that it is beyond the reach of human perception or comprehension. Having persuaded his interlocutor in the *Dialogues on Metaphysics* that "the least motion of matter is able to continue to an infinity of important events, and that each event depends upon an infinity of subordinate causes"—another attempt to master the concept of infinity—Malebranche's representative Theodore asks his interlocutor, "Have you never amused yourself by nursing a caterpillar in a box?" "Oh my," answers his interlocutor. "You all of a sudden go from the great to the small. You are always coming back to insects."[35]

In the spirit of Christian humility, Theodore replies that he is pleased to admire what everyone else scorns. This occasions a complete survey of the insect in all its aspects—metamorphosis, spontaneous generation, *emboîtement*, purposes, and symbolism. The final topic is the transmutation of worms into butterflies. The prophet said, "I am a worm" (Psalms 21). Christ was a worm; man is a worm, but destined for a glorious life. A worm is despicable and helpless. But it closes itself in a tomb from which it is revived. Instead of putrefying, the worm comes to life again in a better form, in a spiritual, winged body. No longer something contemptible, nourished on putrefaction, it "sucks on flowers

---

labyrinth image, has some currency. Jean-Baptiste Duhamel, the reconciler of ancient and modern philosophies, appeals, for instance, to the "seeds hidden in the deep abyss of elements"; Duhamel, *De consensu veteris et novus* (1663), in Roger, *Les sciences de la vie*, p. 333.

[34] Bayle, "Rorarius," in *Dictionnaire historique et critique*, p. 2604. See Leibniz's criticisms of the "double horror" in *Philosophische Schriften* 4:526–27.

[35] Malebranche, *Dialogues on Metaphysics*, p. 257.

magnificently adorned."[36] Once it grew fat, but now it ceases to grow, just as on earth we can earn merit incessantly, whereas in heaven we remain what we are. The butterfly's sexuality and egg-laying propensities in its spiritual form are unproblematic, for its very fertility is an image of the Church after Christ's death. It is apparent, Theodore says, that God wanted to represent the meaning of Christ's existence in the changes of insects, for he made so many of them—"That jumps to the eye."

Many details of this discourse in addition to its general sentiments are borrowed from Swammerdam, who insisted, contradicting the ancients, that it is the same animal and not a new one, born from another type of "egg," that emerges from the chrysalis. In his butterfly chapter, he offered to show the reader how "a poor and wretched insect lose[s] by degrees all motion, and, in appearance, stand[s] consigned to death and the grave; in which seemingly hopeless condition, however, all its former limbs acquire an extraordinary degree of perfection, till, at last, rising from the sepulchre in all the gaiety and magnificence of the richest ornaments, and most resplendent colours, it no longer continues a reptile, creeping upon the earth; but, soaring into the air, changes its slow and heavy pace into the most nimble and unrestrained flights."[37] The butterfly as a symbol of resurrection is of course conventional, and antedates the lens-assisted study of the insect by a long time. What stands out in Malebranche's *Dialogues on Metaphysics*—by contrast with Swammerdam's *History of Insects*—is the mixture of enthusiasm and embarrassment with which the theme is expounded. "I am glad, Aristes," Theodore says, "that this thought came to your mind. For, though it seemed to me most solid, I should not have dared to propose it to you."[38] There is something dubious about it: something unseemly in the comparison of the Savior to a worm, something arbitrary and risky in the search for natural emblems. Malebranche worried that he might be going too far in the direction of insect worship, that physico-theology, in its leap from the sensible to the supersensible, had something in it of the comic genre, in which high subjects and low subjects are absurdly mixed.

Nor is one supposed to draw from the human-caterpillar analogy, though arguably Kircher and Leibniz were both tempted to do exactly this, the conclusion that resurrection is a natural process of development in which the same being will emerge resplendent at the end of its period of latency. Francesco Redi corrected the error of "those who,

---

[36] Ibid., p. 271.
[37] Swammerdam, *Book of Nature*, part 2, p. 10.
[38] Malebranche, *Dialogues on Metaphysics*, p. 273.

from the natural and intelligible changes in bodies, have endeavored to explain the resurrection of the dead." Human beings, Swammerdam had maintained firmly, unlike pupated insects, do not experience an "idle and imaginary death ... or the transformation, as it is called, of their limbs," but a real death, and their resurrection is a miracle surpassing the operations of nature.[39] Redi, who set out to prove that corrupted matter cannot produce new life, saw himself as refuting Kircher's doctrine of palingenesis, his alleged resuscitation of dead flies from their ashes. But Leibniz's teaching on the natural resurrection also violated orthodoxy, in that he made the extraordinary claim that humans and animals alike do not actually die. What we know as death is really the contraction of an organic body and the diminution of perceptual acuity and scope of action to the point where the being itself escapes notice. "Just as animals in general are not completely born in conception or *generation*, moreover, neither do they completely perish in what we call *death*, for it is reasonable that what has no natural beginning also has no ending within the order of nature."[40]

Leibniz naturalized—without removing its wonder—the miraculous, and the motif of physico-theology is the notion of a natural wonder. The rationalist philosophers disagreed strenuously among themselves on many issues—mind-body relations, free will, the nature of substance—but Descartes, Leibniz, Malebranche, and Spinoza shared the conviction that God does not show his presence and power by intervening in and altering a preestablished course of nature. Within their various Christian and anti-Christian agendas, they wanted to invest the ordinary and reliable course of nature with enough of the dramatic to show that it was still stamped by the power and light of God. The regularity, reliability, and universality of natural law must make an impact of sublime inexorability on the imagination. Malebranche's influence here was especially deep, and not only on Leibniz. Vallisnieri admired and frequently referred to Malebranche's statement of the generality of the laws of nature and considered him an observer of the first rank.[41] Nature, for these rationalists, was not simply a regular but faint pulse in the background. It had to give us something extraordinary and almost monstrous if we were to see in it the mind and hand of God. In Malebranche's "Eleventh Dialogue," he shows how the imagination is frightened by the calculations that prove the existence of a thousand small bees inside each adult body and needs reassurance from the reasoning part of the mind. The intellect's appreciation of divine skill and might

[39] Swammerdam, *Book of Nature*, part 1, p. 9.
[40] Leibniz, *The Principles of Nature and of Grace*, no. 6, in *Leibniz*, p. 638.
[41] Vallisnieri, *Opere fisico-medichi* 2:121.

transmutes incredulity and revulsion at the idea of the superpopulation of a single organism into religious awe.

But how, one might wonder, is this different from naturalism? For God might as well not be there. *Emboîtement* is a picture in which God is at rest. Generation is not brought about by powers or faculties, or movements of desire and imagination—not of the creatures, not of God. "All this is done in relation to the laws of motion, laws which are so simple and natural that, although God does everything by means of them in the ordinary course of His Providence, it seems that He affects nothing, that He is involved in nothing, in short that He is at rest."[42] Malebranche's occasionalism is sometimes interpreted to mean that God always acts, that there is no change in movement or perception in the entire universe that is not a direct expression of his choice and will. And yet all discretionary room has been closed off, because God acts by laws that are uniform and invariant. We might as well say "something once acted; now nothing acts." Malebranche's denial of the particularity of divine action, his repeated insistence that monsters, disasters, and misfortunes are all accidental, is as cold as Spinozism, and his God at rest is as antiprovidential a notion as that of a blind, rampaging Nature. It is only in the descent from physics into particularities that something of the warmth of the old cosmological scheme is recovered, so that here religious orthodoxy needed, and was willing to enlist the help of, empirical science.

If the domestic or garden insect was a positive help to physico-theology, the role of the microorganism in the grand scheme of things was less clear. The philosophers knew—though in a remote and theoretical way—that the microworld of corpuscular textures determined the experienced world of qualities and interactions, and they suspected in a vivid way that the microworld of living creatures brought about generation and perhaps death. But how much did they know about the animalcula and their varieties?

The somewhat surprising insistence that the animalcula in general, not just the spermatic variety, were complete animals, I have already noted in Leeuwenhoek. Andry, who quotes almost verbatim from the enthusiast Malebranche, is equally convinced of this: "Animals a thousand times less than a Grain of Dust, have a motion like other Animals; they have Muscles then to move, Tendons, and an infinite number of Fibres in each Muscle; and in fine, Blood or Animal Spirits, very subtle and fine, to fill or move those Muscles, without which they could not transport their Bodies into different places. . . . Our imagination loses

---

[42] Malebranche, *Dialogues on Metaphysics*, p. 253.

itself in this thought, it is amazed at such a strange littleness; but to what purpose would it be to deny it? Reason convinces us of the existence of that which we cannot conceive."[43] This certainty seemed to fade in the following century: the notion of an "organic molecule" as a building block of all living creatures lost its articulation and replaced the "insect" as the smallest living unit. But protozoa and bacteria are not consistently distinguished even in the eighteenth-century literature, with some writers supposing them to be various precursors of fungi, or protean manifestations of a single type of organism.[44] There are many examples of the continuing broad use of the term "worm" to cover insects, amphibians, and reptiles, and in the tenth edition of Linné's *Systema natura* (1758), the protozoa are still identified as members of the class of WORMS, under the order "Zoophytes," and divided further, two editions later, into three genera: *Volvox, Furia,* and *Chaos,* with the infusoria discovered by Leeuwenhoek designated as the single species *Chaos infusorium.*[45] In a much-remarked appendix, Linné made a provisional list of six kinds of "living molecules" that brings to the fore their suspected roles.

1. The *contagion* of eruptive fevers?
2. The *cause* of paroxysmal fevers?
3. The *moist virus* of syphilis?
4. *Leeuwenhoek's* spermatic animalcules?
5. The airy mist *floating in the month of blossoming?*
6. Muenchhausen's *septic agent* of fermentation and putrefaction?[46]

Fertilization, fermentation, contagion, and putrefaction—the probable accomplishments of the animalcula—are here hypothesized as processes involving living entities without any exact grasp of their logic or any precise categorization of the entities involved. Even Pasteur in the 1870s did not distinguish among what he called *végétaux cryptogames microscopiques, animalcules, champignons, infusoires, torulacées, bactéries, vibrioniens,* and *monads.*[47]

Linné's list suggests that the animalcula were already conceived by the mid–eighteenth century as indispensable to the normal operations of nature. This was not the case at the end of the seventeenth, when questions of their origin and purpose could not be avoided but presented difficulties. Were the animalcula products of an original divine creation, and simply the invisible agents of God, or were they continu-

---

[43] Andry, *Breeding of Worms,* p. 189.
[44] Bulloch, *History of Bacteriology,* chap. 8; Dobell, *Antony van Leeuwenhoek,* p. 378.
[45] Dobell, *Antony van Leeuwenhoek,* p. 377.
[46] Ibid.
[47] See Bulloch, *History of Bacteriology,* p. 187.

ously produced? The changing fortunes of the theory of spontaneous generation prove that the theories handed down from antiquity had not always been unacceptable to Christians. Spontaneous generation became a theologically significant issue only as a nontranscendent origin for life became an openly discussed possibility, and as the special creation of each species became a bastion of defense for theism. When Saint Augustine, for example, speculated that all creatures are produced from tiny seeds lying quietly in the earth until the time comes for their germination, he evidently did not believe that these seeds were necessarily produced by parents of the same kind or through sexual congress and fertilization.

The vagueness of the philosophical notion of "production" made equivocal generation—generation from something unlike the thing generated—a self-evident possibility, indeed something of a tautology. As Cardano says, "generation is always produced by something else." Grains are produced from earth and moisture, animals from seed and blood or eggs, ashes from wood, "nor is there anything so tiny that it is produced from nothing." Corruption is simply another kind of transformation and is equally spontaneous. An apple decays and is changed into worms, water is changed into steam by the heat of the sun, wood is changed into ashes; indeed, it seems that one may speak in the latter two cases of generation or corruption indifferently. What remains constant is a primal matter, which, in the language of Aristotle, assumes different forms.[48] In keeping with the general Aristotelian tendency to suppose that invariable and regular effects are not susceptible of specific explanation, insofar as they form part of the regular course of nature, these qualitative transformations do not call for scientific investigation except where there is something anomalous about the process.

Although it is customary to do so, it is perhaps overinterpreting to refer to the Aristotelian doctrine of equivocal generation by the nineteenth-century name of spontaneous generation, as though a positive theory about the conversion of nonliving filth and stagnant fluids into living organisms were implied. The distinction between the living and the nonliving up to the beginning of the eighteenth century was not sharp enough to support such an interpretation. "After lying for some time," Aristotle says, "snow turns slightly red, hence the larvae it produces are red too, and hairy."[49] I have already noted his belief that humans produce their own lice and fleas. Although some insects, such as spiders, grasshoppers, cicadas, and locusts, are produced from animals like themselves, he says, "Some are not produced from animals at all, but spontaneously from dew, during spring, warm spells in winter,

[48] Cardano, *De subtilitate*, p. 83.
[49] Aristotle, *History of Animals* 552b.

from putrefying mud and dung, in wood, in the hair of animals, in their flesh, in residues."[50] But we do not know in what sense we should understand this notion of "production"; we can only infer that Aristotle does not hold the general thesis that insects are sexually dimorphic and reproduce their kind by mating and laying eggs.

The ability of nature to bring forth was always emphasized in Renaissance nature philosophy. Cardano discussed the emergence of life from putrefaction, as well as the regenerative abilities of serpents, shrimp, swallows (which can grow new eyes), chameleons, and tortoises.[51] According to Moffett, most insects can be produced directly in a "first generation," as when bees are produced from the carcass of a rotting bull, ox, cow, or calf, or through sexual mating and the laying of eggs. We see both the productive power of corruption, following from the first human sin, and "Nature acting successively in a circle, and constantly by a perpetual motion running back into her self," in the cycle of egg-worm-chrysalis-butterfly-egg.[52] In the case of fleas, for example, "Their first Originall is from dust, chiefly that which is moistened with man's or Goats urine. Also they breed amongst Dog's hair, from a fat humour petrified, as Scaliger affirms. A little corruption will breed them, and the place of their originall is dry filth."[53] It is manifest, he says, that spiders are "bred of some areall seeds putrefied, from filth and corruption, because that the newest houses the first day they are whited will have both Spiders and Cobwebs in them," but these then propagate by copulation and the laying of eggs.[54] Although his predecessor Pennius derided Aristotle for this view, "yet all Philosophers with one consent agree that the more imperfect small creatures are bred of dew."[55] The dew and sun are the major cause of caterpillars, which also hatch from eggs. Moffett is liberal and generous and avoids dogmatism: "I should maintain that they are not bred only one way, but all these waies."[56] Only it must be false, he says, that out of the corruption of every creature, whether bull, lion, or some other animal, a new kind of insect is produced, for this would make generation infinite and is contrary to experience.

Kepler too portrays a sort of general life force, present in the earth and multiply differentiable. "All plants," he says, "are offspring of one and the same universal principle, inherent in the earth and related to plants as the principle of water is to fish, of the human body to lice, of

[50] Ibid., 550b.
[51] Cardano, *De subtilitate*, book 9, in *Opera omnia* 3:507ff.
[52] Moffett, *Theater of Insects*, epistle dedicatory.
[53] Ibid., p. 1102.
[54] Ibid., p. 1072.
[55] Ibid., p. 1040.
[56] Ibid.

the bodies of dogs to fleas, and of sheep to some other kind of louse. Not all plants anyhow originated from seed, but most of them arose spontaneously, although they have propagated themselves since by seeding."[57] For Kepler, these formative principles are distributed through the earth, water, and air. We see them at work not only in the emergence of plants and animals, but in the formation of snowflakes and flowers of frost, in the rhomboid cubes in sulfates of metals and crystals of saltpeter, and in the patterns of honeycombs, flower petals, and pomegranates. There is no worry here about a dangerous ascription of autonomy or intelligence to nature, such as can be found in the late-seventeenth-century attack on hylomorphic principles and plastic natures, for all this is said to happen under a vaguely theistic umbrella, from a principle "not to be sure as discursive reasoning discovers, but such as existed from the first in the Creator's design and is preserved from that origin to this day."[58] Harvey was less than definite on the question whether the generation of lower beings is always cyclical; the principle that all animals come from eggs does not seem to imply for him that all eggs are laid by parents of the same kind. Commenting on Aristotle's claim that all creatures arise from a vegetal primordium either by chance and spontaneously, or from something preexistent such as an egg or blood, Harvey stated that "there are only two kinds of productions, inasmuch as all animals produce another animal either actually or potentially."[59] This can be read as excluding chance productions, but it does not show a firm commitment to generation lines strictly divided by species. Observation and experience, then, tended to confirm the existence of equivocal generation and generation from putrefaction, while the arguments against it were a priori, versions of the principle that effects cannot be superior to causes.

As Monconys reported and its *Journal Books*, which recorded the proceedings of its weekly meetings, confirm, the newly established Royal Society meeting at Gresham College was already conducting experiments on the generation of insects in closed vessels in 1662 and arriving at negative conclusions. An entry for 15 October shows that the members drew the inference that, if all creatures arise from similar parents, the air must be full of invisible drifting semina. The finding of some parasitic worms in a salmon, it is reported, gave occasion

of discoursing upon Equivocall generation; and it was considered whether all Animals, as well Vermin and Insects, as others, are produced by certain seminal Principles, determined to bring forth such, and no other kinds, some con-

[57] Kepler, *Six-Pointed Snowflake*, p. 35.
[58] Ibid., p. 33.
[59] Harvey, *Disputations*, p. 326.

ceiving, that where the Animal itself does not immediately furnish the seed, there may be such seeds, or something Analogous to them, dispersed through the Air, and conveyed to such matter, as is fitt and disposed to ferment with it, for the production of this or that kind of Animal; others observing that those Semina said to be carryd in the Air, would have their Seminal Virtue impaired, if not destroyed, by it.[60]

Drs. Glisson and Charleton were assigned to monitor some experiments; they found that after ten weeks the putrefying blood, flesh, and brains of animals in a glass covered with cotton produced neither worms nor anything else.[61]

In the *Micrographia*, however, Hooke does not rule out the possibility. "We cannot presently positively say," he argues, "there are no animal substances, [produced] either mediately, as by the soil . . . of the plant from whence they sprung, or more mediately by the real mixture or composition of such substances. . . . In such places where such kind of *putrefying* or *fermenting* bodies are, [they] may, by a certain instrument of nature, eject some sort of seminal principle which . . . may produce various kinds of insects or Animate bodies."[62] Mold and mushrooms, he says, require no seminal principles, but spring directly from putrefying flesh or vegetable matter; they are like crystals in that their form can arise through ordinary laws, as one can see from the mushroom shape of the cloud of an extinguished candle.[63] The possibility of equivocal generation was important to the program of the *Micrographia*. Hooke envisioned an orderly and unified science of morphogenesis, which would explain the growth of complex animal forms by beginning with the basic structures of nature and showing how they arise and how the more complex forms develop out of them. The possibility of systematization on the basis of empirical observation seemed to require that there be no sharp divisions among the animal, vegetable, and mineral kingdoms. Kenelm Digby, the necessitarian, had said around 1644 that plants were not living things but only chemical compositions involving mainly a nitrous salt. He found experimentally that when he sowed barley seeds outside a villa in Rome, stalks of pure saltpeter an inch or two high sprang up the next morning.[64]

The thesis that Francesco Redi set out to refute in his *Experiments on the Generation of Insects* (1668) was a moderate one: new life cannot

[60] Birch, *History of the Royal Society* 1:117.
[61] Monconys, *Journal des voyages* 2:27.
[62] Hooke, *Micrographia*, p. 123.
[63] Ibid., pp. 127–28.
[64] Digby, *Of Bodies*, in *Two Treatises*, pp. 61ff.

come from decay and putrefaction. This was not the same as a denial of
spontaneous generation or of forces in nature capable of producing a
new animal. Redi accepted and even embraced the thesis of the production of animal life from a vigorous vegetable soul. His opinions set him,
he said, against the received view of the schools, which maintained that
the earth, whether or not it once had the power to generate large creatures (the Stoics had believed that human beings sprang from earth), is
now so exhausted that it can produce only lesser ones such as flies,
wasps, scorpions, and ants.

His reasons for taking this stand emerge more clearly when one considers his opponents: Kircher, who believed that new flies could be re-
generated in the dead bodies of old flies, and Digby, who made tantaliz-
ing suggestions about palingenesis in his daring *Treatise of Vegetation*,
and who, in his *Treatise of Bodies*, dealt with the reproduction of plants
from their ashes. In Redi's view this whole approach was wrong: "The
Holy mysteries of our faith," he says, "cannot be comprehended by
human intelligence; unlike natural things, these are of the special work-
manship of God, who is believed to be omnipotent, and therefore it is
possible to believe blindly in all his works, for so they are best understood."[65]

Kircher, who, like Digby, was never regarded as entirely trustworthy,
had described various experiments in his plague treatise and later in his
Mundus subterraneus (1665) involving the generation of microscopic
and macroscopic animals. Take a phial of water, he says, mix it with
earth, add some pieces of chopped serpent, and expose it to the air.
Little balls will appear, which will turn into worms, and these will then
develop wings. Kircher did not distinguish between the hypothesis that
these creatures appear when drifting animate seeds settle into a com-
fortable nidus and develop and the hypothesis that they are created
from decaying matter through the action of a form-producing force; his
emphasis is simply on the cause-effect relations among the corruption of
internal humors, the poisons of plants, the breath of malignant animals
such as serpents, toads, and newts, rotten foods, and the production of
worms and insects.

Redi decided to firm up the experimental apparatus. Through a series
of trials with covered and uncovered dishes of meats and other foods,
including ox, deer, buffalo, llama, tiger, lamb, dog, kid, rabbit, fish,
milk, fruits, and vegetables, he discovered that these did not produce
worms unless left uncovered to the air.[66] But the rest of his treatise is

[65] Redi, *Experiments*, p. 36.
[66] Ibid., pp. 75–76.

taken up with a defense of the productive power of the vegetative soul. Here Redi criticizes Father Liceti, the author of several treatises on generation, who asserted on a priori grounds that the vegetative soul could not produce a sentient animal soul, for lower forms cannot produce higher; accidents such as warmth, light, and stellar rays cannot produce substances, and general causes cannot produce particular things. The generation of animal organisms must therefore proceed, Liceti said, either from the remnants of the animal soul in decaying matter or animal excrement, or from living atoms that fly about through the air. While noting that others too, such as Gassendi and Joachim Jungius, are of the opinion that animals always come from eggs, Redi insists that another mode of generation may be found "in the peculiar potency of that soul or principle which creates the flowers and fruits of living plants, and is the same that produces the worms of these plants. Who knows?"[67] Indeed, he says, the production of worms and winged animals may be the primary purpose of the plant, the fruit only secondary. There is nothing degrading or paradoxical in the production of ugly worms from beautiful shrubs: the worm is not an intrinsically lowly creature. Rather, "low and high are unknown terms to Nature, invented to suit the beliefs of this or that sect, according to the needs of the case."[68]

The strong thesis that no insect or other animal can be produced except by eggs laid by parents of the same kind thus had some a priori backing, but artificial observations and experiments were needed to counterbalance the weight of common sense and tradition.[69] Defenders of the strong thesis pointed out that the insect body, even the worm body, was as complex and differentiated as that of the quadruped. It was not a soft blob similar to the decaying material or dirt of which it was allegedly made. Also, the microscope had shown sexual organs and even unlaid eggs in numerous species of insects. Leeuwenhoek was definite on the subject. "All animals were created at the beginning of the world," he says, "and no new species have since been produced. . . . It is as possible for the Rocks of Stone to produce Horses as for any animal whatsoever to be engendered of putrefaction."[70] He made specific studies of twenty-six insect species, showing that, in each case, parents

---

[67] Ibid., p. 91.

[68] Ibid., p. 94. This does not prevent him from describing some of the worms he finds in the head of a sheep as "disgusting" (p. 117). He compares lovesickness with parasitic infection: "And deeply gnaws the tissue of the brain / Thus does love madden with excess of pain" (p. 119).

[69] Schrader, *Microscopiorum usu*, pp. 10ff.

[70] Quoted in Cole, "Leeuwenhoek's Zoological Researches," p. 3.

and eggs could be demonstrated.[71] The insects inside fruits, galls, and weevils in grain got there, he found, by boring or chewing holes and by the deposition of eggs through the narrow, pointed ovipositor of the female. He showed that aphid bodies too were subject to invasion by other winged organisms that laid eggs there.[72]

Yet there was still doubt, even after the turn of the century, about the necessity of similar parents. The philosophical principle that organization cannot emerge from chaos, which he claimed common people still failed to grasp, was enunciated by Malebranche, who recounted what was by then a standardized demonstration in the *Dialogues on Metaphysics*. Theotimus takes a small piece of meat, encloses it in a bottle, and covers it with a piece of crape. Flies lay their eggs on the crape, worms gnaw through it and reach the meat, which they devour, and, when the whole thing begins to smell, Theotimus throws it away. Hermetic sealing of the vessels does not produce flies or larvae, and it cannot: "What is more incomprehensible than that an animal should be formed naturally from a little rotten meat? It is infinitely easier to conceive of a piece of rusty iron changing into a perfectly good watch" as it is to imagine a mouse constructed from debris, "for there are infinitely more parts that are more delicate in a mouse than in the most complete clock."[73] And mechanical laws, especially the basic law that a body moves in the direction of the least pressure, are more fitted to destroy machines than to construct them.[74] But his assumption that animalcula in open vessels always came from eggs released by flying animalcula was not borne out by an experiment presumably performed by him in 1707. Only inoculated vessels produced large numbers of animalcula, and he noticed certain conjoined pairs of animalcula in his vessels, which he at first assumed to be fighting, but later concluded to be mating.[75]

The propagation of plants from slips and cuttings proved that new individuals could come into being by means other than growth from seeds, and, by 1739, it was known that there were analogous phenomena in the animal kingdom. Not only was regeneration (for example, of missing limbs) possible, but so was the production of a complete new organism such as a polyp by budding.[76] Plastic forces reentered natural

---

[71] Ibid., p. 7. Cole notes two failures, the freshwater eel and the marine mussel.

[72] Ibid., p. 5.

[73] Malebranche, *Dialogues on Metaphysics*, p. 263.

[74] Ibid.

[75] *Historie de l'Academie des Sciences* (1707), pp. 8–9, in Malebranche, *Oeuvres complètes* 19:771.

[76] Trembley's earlier studies, which were described definitively in his *Mémoires pour servir a l'historie d'un genre de polypes d'eau douce* (1744), were reproduced by Henry

philosophy with a vengeance. *Ex ovo omnia* had been a good conjecture, but it now looked to be false. One could only hypothesize with good reason that every living thing comes from similar parents: first, as Swammerdam had argued, metamorphosis was imaginary, and the stages of insect development were stages in the life of a single individual being. Second, as Redi had argued, no living thing comes from putrefaction. Third, as Leeuwenhoek had argued, all parasites are exogenous to the bodies they infest.

The animalcula of Leeuwenhoek, however, threatened to destroy these generalizations.[77] Hearing of the work of Redi and of the "divers opinions expressed concerning the generation of little animals," Leeuwenhoek decided to carry out his own trials. He took two glass tubes, filled them with pounded pepper and clean rainwater collected "in a clean china dish (in which no victuals had been put for quite ten years)," drew the tops of the tubes up to a small point, and sealed the top of one a quarter of an hour later, leaving the other open. The results were that the open glass showed animalcula after three days. The closed glass did not at first reveal any animalcula when examined from the outside, but, when it was broken open, "I perceived in it a kind of living animalcules that were round, and bigger than the biggest sort that I have said were in the other water, though they were yet so small that it was not possible for me to discern them through the thickness of the glass tube." He concluded that life, or some forms of it, could, after all, appear in closed vessels. "I bethought me that when that Gentleman aforesaid spake of living creatures (*dierkens*) he meant only worms or maggots, which you commonly see in rotten meat, and which ordinarily proceed from the eggs of flies, and which are so big that we have no need of a good microscope to descry them."[78] The question whether life could appear in closed vessels blurred three distinct questions: first, whether air was needed to support life; second, whether all animals came from eggs laid at some time by parents; third, whether eggs or seeds could survive for indefinitely long periods while appearing inert and dead. Leeuwenhoek was evidently not prepared after all to recognize the emergence of living beings from chaos, recognizing rather that there must be seeds in rain or elsewhere that were able to develop even in the absence of air into organisms.[79]

Louis Joblot carried out further investigations into these questions before 1720, concluding philosophically that "division and separation

Baker, who published his *Attempt towards a Natural History of the Polype* in 1739.

[77] See Ruestow, "Leeuwenhoek and the Campaign."

[78] Letter 32, 14 June 1680, to Thomas Gale, in Dobell, *Antony van Leeuwenhoek*, p. 198.

[79] Ibid., p. 199.

of parts cannot create a being which swims, seeks food and reproduces, an intentional being," and that the experiments of several competent philosophers had shown that "all animals, of whatever nature, come from eggs."[80] He pointed out that bad-smelling, putrefied infusions actually contained fewer animalcula than fresh ones, that corrupted blood, though it smelled insufferable, produced no animalcula after nearly a month in summer, and that one saw various animalcula succeed each other in the same infusion over a course of thirteen to fourteen months.[81] "These experiences seem to me of a sufficient quantity to show that neither alteration nor corruption nor bad odor are the cause of the generation of animals." He explained the presence of animalcula in infusions by the hypothesis that "there fly or swim in the neighboring air an innumerable number of very small animals which, applying themselves to the plants which suit them, pause and there take on some nourishment and deposit their young, while others lay eggs there in which new insects are contained."[82] These insects are perhaps attracted by the smell of the infusions or stimulated by spiritous corpuscles emanating from the plants, which provide a good habitat for them.

The hardiness of life forms that can persist in small spores, survive boiling, live without oxygen, penetrate corks and other seals, and reproduce by fission provided a continual challenge through the eighteenth and nineteenth centuries to opponents of spontaneous generation. The experimental evidence—and the breakdown of the earlier assumption that microorganisms were fully articulated insects with limbs and organs—tended to support it, while only dogmatists felt safe in ruling against it. John Turberville Needham's infusions swarmed with moving particles even in corked-up and long-heated solutions, and forced him to conclude that a vegetative force was dispersed throughout matter.[83] Meanwhile, before putrefaction was interpreted as an effect of infection by microorganisms, its logic was hard to unravel. If putrefaction was a degraded state of a material, how did it provide a nidus for complex living creatures? If "exalted" substances like camphor had antiseptic properties, did this not imply paradoxically that they were both hostile to life and preservative of it? A little salt or oil seemed to favor the growth of microorganisms, while too much put an end to it. The situation was not much clearer half a century later: Edward Wright in the *Philosophical Transactions* of 1756 tried to systematize these confusing

[80] Joblot, *Descriptions et usages*, p. 44.

[81] Ibid., p. 45.

[82] Ibid., p. 46.

[83] Sloan, in "Organic Molecules Revisited," argues that organic molecules were not an example of "micrographia illusoria" but actually particles in Brownian motion.

relations and arrived at the conclusion that a degree of "exhaltation" of oils and salts nourishes plants and animals, more favors putrefaction and the generation of animalcula, more still the destruction of all life.[84] The discoveries of Pasteur and others that fermentation is a "disease" of mash, and that putrefaction implies invasion by living organisms, finally make the logic clear: putrefaction cannot be the cause of animal life because it is the effect.

In summary, the denial of generation by chance was not only a fixture of the orthodox doctrine of special creation in its adversarial relationship to the pagan doctrine of the productive powers of nature. It also exemplified the uniformity of nature—nature as artifact—and the economy of means in its production. This structural support of rationalist metaphysics seemed to show itself not simply in the systematicness and closure of the laws of mechanics, but in the same systematicness and closure in the world of living creatures. The development of a new logic of identity in the refutation of Cardano's claim that generation is essentially the production of something unlike the producer helped shape precisely the notion of a living organism while serving an apologetic function in the context of the Christian theory of immortality.

Accordingly, one must regard with some caution forceful claims about the absence of biology in this period. The discipline is not named as such, or institutionalized as such, and it is correct to say that biology had no subject matter before 1750 in view of the lack of consensus about what was alive and what life's defining characteristics were. Yet the field was defined implicitly, insofar as it was the data of early microscopy that contributed, far more than physics and astronomy in the same period, to the construction of a worldview. In the absence of any formal conception of biology, the microscope taught the unity of nature.

Plants and animals, the microscope showed, have much in common when it comes to generation. In both kingdoms, mating produces a fertile seed, egg, or fetus. This point was stated by Grew in 1676 and again in his *Anatomy of Plants* of 1682, and confirmed by Malpighi. It had not been apparent in the mid–seventeenth century that almost all plants have male and female organs and propagate from seeds or spores. When Sir Thomas Browne expressed the wish that "we might procreate like trees,"[85] he meant specifically to rule out the conjunction of male and female. The "maleness" and "femaleness" of palm and holly trees had long been recognized, but these concepts were, in the absence of a

---

[84] Wright, "Microscopical Observations," p. 557.
[85] See Gunther, *Early Science in Oxford* 3:161. Browne later married happily.

specific theory of generation, only vaguely analogous to the notions as they applied to animals. The conviction that plants and animals were both organic systems of some complexity was taken briefly to extremes. As Grew said in his preface to *The Anatomy of Plants* (1682),

> Your Majesty will here see that there are those things within a Plant, little less admirable, than within an Animal. That a Plant, as well as an Animal, is composed of several Organical Parts; some whereof may be called its Bowels. That every Plant hath Bowels of divers kinds, conteining divers kinds of Liquors. That even a Plant lives partly upon Aer; for the reception whereof, it hath those Parts which are answerable to Lungs. So that a plant is, as it were, an Animal in Quires; as an Animal is a Plant, or rather several Plants bound up into one Volume.

The fibers of hardwood plants, he says, are "so closely couched and drawn up together, as to lye rather after the Manner of the Vessels in the Liver, Testicles, Glands and other Viscera in Animals."[86]

Swammerdam's and Malpighi's views were similar, and the idea that plants and animals are like each other and distinct from the rest of nature had a galvanizing effect on the metaphysics of Leibniz, who referred, in one of the letters expounding his philosophy of 1686 to the Cartesian theologian Antoine Arnauld, to the "wonderful analogies which have lately been discovered"—a reference presumably to Swammerdam's statement that "the same changes . . . which we observe in vegetative animals, are equally observed in sensitive ones, so as to afford us in all God's works the most manifest proofs of his infinite wisdom and power . . . for as the foundations of all created beings are few and simple, so the agreement between them is most regular and harmonious."[87] Leibniz was impressed not only by talk of harmony and regularity but by the implication that the distribution of souls in nature was far greater than the Cartesians had realized. In view of their organic complexity, we cannot deny a soul, and so a "real unity," to every plant, even while recognizing that this soul plays no active role in its growth or vitality. And preformation, which he learned from Malebranche, induced in him the experience of one of those wonderful confluences of ideas that enables a philosopher to say "my system." Against the pansophical hypothesis that we bear within us the seeds of knowledge, Leibniz laid preformation and found that it fit. As the animal unfolds itself into perfection, so too does the whole world, and the individual mind.[88]

---

[86] According to Leeuwenhoek, "Concerning the Texture of Trees," p. 658.

[87] Swammerdam, *Book of Nature*, part 1, p. 9.

[88] Cf. Comenius, "It is not necessary that anything be brought to a man from without,

Leibniz's particular brand of pananimism, which denies the existence of a world soul or the spiritual information of matter, and instead fills the world with a plenitude of visible and invisible creatures, rested on the discovery of minute insects and animalcula. The influence of Malebranche's chapter in *The Search after Truth* (1675) on the errors of the senses and their inadequacy with respect to the very small, and on the microscope's revelation of "living atoms,"[89] is not to be underestimated. Interested early on in Kircher's microscopical worms, Leibniz followed the micrographical literature with attention and acquired his own copy of the *Micrographia* in 1678. A correspondent wrote to him in the same year to say that Huygens had brought some microscopes back from Holland with which one could see an infinity of little animals in a drop of water, "which has made people conclude that everything in nature is animated."[90] Power had said much the same thing in his *Experimental Philosophy*: "It is not to tell in what a small particle of Matter, life may actually consist and exercise all the functions too of Vegetation, Sensation, and Motion: so that *Omnia sunt animarum plenum*, may have more truth in it, than he could either think or dream of that first proposed it."[91]

Leibniz believed, from at least 1686 onward, that even apparently inert objects were inhabited by multitudes of animalcula; at times he seemed to believe that material objects were not simply inhabited by, but actually constituted by, dense swarms of tightly packed microorganisms.[92] The direct inspiration for this view may well have been Fontenelle's *Conversations on the Plurality of Worlds*, the first edition of which appeared in 1686. The book is concerned mostly with telescopic discoveries and with an explanation of the Copernican system, but a natural transition to the microscope is made under the key word "population." It is not difficult, Fontenelle's natural philosopher says, to believe that other planets are as densely inhabited as the earth when our world is full of invisible living beings.

We see from the elephant down to the mite; there our sight ends. But beyond the mite an infinite multitude of animals begins for which the mite is an elephant, and which can't be perceived with ordinary eyesight. We've seen with lenses many liquids filled with little animals that one would never have

---

but only that which he possesses rolled up within himself be unfolded and disclosed, and the stress be laid on each separate element." *Great Didactic*, p. 194.

[89] Malebranche, *Search after Truth*, pp. 79–93.

[90] Justel, letter to Leibniz, 22 July 1678, in Leibniz, *Sämtliche Schriften und Briefen*, series 1, 2:354.

[91] Power, *Experimental Philosophy*, p. 23.

[92] Garber, "Leibniz," p. 37.

suspected live there, and there's some indication that the taste they provide for our senses comes from the stings these little animals make on the tongue and the palate. . . .

Many bodies that appear solid are nothing but a mass of these imperceptible animals, who find enough freedom of movement there as is necessary for them. A tree leaf is a little world inhabited by invisible worms, and it seems to them a vast expanse where they learn of mountains and abysses. . . . Even in very hard kinds of rock we've found innumerable small worms, living in imperceptible gaps and feeding themselves by gnawing on the substance of the stone.[93]

In 1687 Leibniz argued much the same way:

Our experience is in favor of this great number of living things; we find that there is a prodigious quantity of them in a drop of water tinctured with powder and with one blow millions of them can be killed so that neither the frogs of the Egyptians nor the quails of the Israelites . . . at all approach this number. . . .

. . . . [A]s nothing is so solid that it has not a certain degree of fluidity, perhaps the block of marble itself is only a mass of an infinite number of living bodies like a lake full of fish, although such living bodies can be ordinarily distinguished by the eye only when the body is partially decayed.[94]

Putrefying substances evidently do not produce worms but only enable them to become visible!

So long as the animalcula were still regarded as insects and complete animals, the old valorizing strategies of physico-theology could be called into play. The insect provided no counterexample to natural theology and plenty of reinforcing material. It was a fellow creature, part of the economy of nature, and, to some, it was still a living symbol of the virtues. And if the view that the cosmos is a harmony needed correction by the blunt recognition that its parts might be at war, it was still possible to maintain that conflict was part of the harmony.[95] A metaphysics like Leibniz's, which gives an absolute value to plenitude rather

[93] Fontenelle, *Conversations*, p. 45.

[94] Leibniz, letter to Arnauld, 30 April 1687, in *Discourse on Metaphysics*, pp. 194–96.

[95] Thus Rösel von Rosenhof, the talented illustrator of numerous insects and water zoophytes, points out that "[i]nsects ruin our plants, our trees, our fruits; they not only ravage our fields and gardens, they attack our possessions, our clothes, our furs, our meal, they bore through our wood, our walls, even us. So someone would perform a great service who examined these insects and learned how to destroy them. . . . the enjoyment I had in the observation of their cunning did not distress me or prevent me from looking for powerful ways to kill them." *Die monatlich-herausgegeben Insecten-Belustigungen*, part 1, 1:2.

than convenience, may coexist with determined practical efforts to rid oneself of particular irritants. In his *Insecto-theologia* of 1738, Lesser argued, like John Ray before him, that the purpose of the insect is to inculcate virtue, teaching people to wash their underclothes, clean their houses, and shake out their cloaks, and in general to make them prudent, diligent, and clever. The picture that was to emerge in the nineteenth century, and later in the era of monoculture and the chemical armament of the farmer, of nature as a theater of blind proliferation and vicious competition is restrained by the focus on adaptation and harmonious coexistence.

When it came to the animalcula, however, the physico-theological formulas were sorely tested. The special creation of each variety of animalculum was not regarded as a problem before their types had been distinguished and classified. And no one to my knowledge discusses the whereabouts of the animalcula during the Flood, or the problems that would have been involved in marshaling and loading them. This is somewhat surprising, because as late as 1668, John Wilkins was proving mathematically that the ark of Noah was sufficient, with its given biblical dimensions, to house the number of known species, indicating that learned fundamentalism was still well established. But their existence tended to push the entire idea of a creation under the control of a single being—one who marks the sparrow's fall—into absurdity. And, despite generous assessments of their organic complexity, these creatures were after all not very much like animals and the moral lesson was not readily available: they did not exhibit sagacity, practicality, maternal care, social organization, or anything similar. They moved, swam, ate, and appeared and disappeared as though no one cared about them at all.

Eventually, even these aspects of the microworld might be put to use. The naturalist John Hill saw minute natural history as an antidote to involutional melancholia and a corrective to anthropocentrism. "[A]lthough we may suffer Pride to persuade us that all Things are made for our Purposes and Pleasure, Reason contradicts it," he concluded in 1752. One might even say that it was to humble us that God placed this world out of our reach, but this too would be to suppose anyone cares what benefits humans derive from anything. The world of microorganisms is simply a spectacle from which the scientific observer stands apart. The Stoics looked at the puppet show of ordinary human life, held back from it, and yet found solace in its contemplation; the modern naturalist's engagement with it was semi-ironical and the spectacle alternately cheerful and appalling. A century before Darwin and Tennyson, nature was beginning to look to Hill red in tooth and claw. Walking in the garden, he heard a "strange fluttering of wings" and looked down to see an ichneumon wasp descending on a caterpillar, which showed its terror at the approach of the fly by twisting, rolling, and writhing. He could see no reason for such wanton cruelty, he said, or why, in the disposition of nature, the eggs of one species must be hatched in the flesh of another. "Various are the arts of death," he ruminates, "not only from the mischievous Inventions of Men, but from the Provision of Nature for its several Productions." Looking through the microscope, he was astonished by the feeding behavior of his zoophytes: "such a science of Butchery, so universal, so varied, and so hurried in all its parts, human cruelty itself never offered." His animalcula experience "agonies of hunger and disappointment"; the contractions of their bodies after unsuccessful food searches "shewed their distaste and Surprize at finding nothing but Water in their Mouths instead of Food." Yet an atmosphere of fun prevailed overall in the microworld, with many of its denizens seeming to enjoy themselves "with great Jollity, scudding from Part to Part with utmost Agility, rolling and turning themselves over at Pleasure."[96]

Does not science promote libertinism? If there is a moral lesson in the animalcula it would have to be: carpe diem. But physico-theology need not be human centered, asserting that the provisions of nature are all for us. It may be creature centered, in which case the provisions of nature for each particular species are what prove divine beneficence. As Boyle put it, a little shakily, we have to admire "the consideration of the vastness, beauty, and regular motions, of the heavenly bodies; the excellent structure of animals and plants; besides a multitude of other phaenomena of nature, and the subserviency of most of these to man."[97] Some things are just there, but not specially for our benefit and enjoyment. Kant still needed to criticize, in the *Critique of Judgment* (1790), the old anthropocentric forms and the theory of "uses," reformulating the notion that nature seems to be made not only for itself but for us as an element of aesthetic as opposed to cognitive judgment.[98]

Indeed, the reference to uses in natural history surveys that detail the therapeutic and economic significance of a species to human beings survives, along with occasional references to their symbolic significance, into and beyond the period of classical microscopy. However, the formula wavers. Consider once again John Hill, who expresses most of the paradoxes of this form: its crude utilitarianism blended with a call for transcendental wonder; its anthropocentrism and its conception of na-

---

[96] Hill, *Essays*, p. 256.
[97] Ibid., p. 515.
[98] Kant, *Kritik der Urteilskraft*, in *Gesammelte Schriften* 5:359ff.

ture as a self-contained system; its justification of objective science as a form of worship:

> Even what the vulgar call the most abject things will shew a wonderful utility; and lead the mind in pious contemplation higher than the stars. The poorest moss that is trampled under foot, has its important uses: is it at the bottom of a wood we find it? why there it shelters the fallen seeds; hides them from birds, and covers them from frost; and thus becomes the foster father of another forest! creeps it along the surface of a rock? even there its good is infinite! its small roots run into the stone, and the rains make their way after them; the moss having lived its time dies; it rots and with the mouldered fragments of the stone forms earth; wherein, after a few successions, useful plants may grow, and feed more useful cattle![99]

Grass! toadstools! beetles! All contribute to the beauty and economy of nature.

Though Hill is thinking of the interests of flesh- and milk-hungry humans, there is a marked revision of the values of high and low here, an awareness that all creatures are part of a single system in which nothing can be scorned. Science had originally dealt with noble objects. In the *Parts of Animals*, Aristotle, though perhaps a better natural historian than he was an astronomer, writes movingly of the difficulties associated with studies of celestial phenomena by contrast with that of sublunary things subject to generation and decay. The former are precious and divine but scarcely accessible to us, "whereas respecting plants and animals, growing up among them as we do, we have abundant information and ample data may be collected concerning their various kinds if we are willing to take sufficient pains." Yet although "both departments have their charms," knowledge of things above us is, he says, more precious and ultimately gives more pleasure than all our knowledge of the world in which we live, "just as a casual half-glimpse of persons we love is more delightful than other things."[100] The objects of the microscope are remote and somewhat difficult of access, but they are not high subjects. The instrument brings a sophisticated beauty into the world, in the regular and symmetrical forms and patterns it makes visible; but its subjects were from the point of view of the ancients demeaned and demeaning: the products of putrefaction, scrapings from the teeth, ejaculates and ejecta; eggs, feces, phlegm and pus, hair,

---

[99] Hill, *Hypochondriasis*, p. 27.

[100] Aristotle, *Parts of Animals* 644b23–645. See Carlo Ginzburg's discussion of the title page of the edition of Leeuwenhoek's *Epistolae* published in Leyden in 1719, which shows a man ascending a mountain toward a cornucopia, assisted by Father Time. Ginzburg, "High and Low," p. 40.

feathers, and dust. Leeuwenhoek noted how interesting his urine looked
after he had eaten asparagus. "I saw very prettily all its little tubes."
The vigorous crudity of his letters on hen's feces and his own diar-
rhea—one face of the indisputable objectivity of his science—is an example for the historian to marvel at.[101]

As Wolfgang Philipp remarks in his great study of eighteenth-century
apologetics, physico-theology is unattractive to the contemporary reader.
"The style makes a repellent impression, loaded with affect, blowsy,
tending to sentimental diminutives, missionary, urging, emotional, then
exact and fussy, occasionally inadvertently funny, trivial, and utilitarian."[102] With its rhymes and biblical citations, it reads as trite and
unreflective. We are apt to find in this attention to small objects a kind
of bourgeois triviality, like that of a collection of knickknacks on a
mantelpiece, which the ubiquitous cabinet in a sense really was. The
expansive rhetoric of the spectacle and harmony of nature produces the
impression of a kind of scientific kitsch. This illustrates the difficulty of
being profound without being vulgar: physico-theology tended to be
both. It invited its own undermining first by supplying to the theory of
natural selection the data of adaptation, which Darwin was to recon-
ceive as the negative result of variation and competition rather than the
positive result of fabrication, but also in its portrayal of nature as spectacle. Physico-theology focused attention on structure and function and
extracted the animal from the field of history, morals, and citations: it
controlled its referentiality even as it sought to control the implications
of that extraction through theological reference.

In summary, there were two facets to the objectification of nature.
Exact natural history encouraged its contemplation as a system: as forming in itself a unity, a plenitude, and a harmony, independent of human interests. It also encouraged the subtraction of moral attributes
and purposes: the new discourse of nature was as impoverished in cultural terms as it was enriched in observational terms. Urbanization and
mechanization, considered as sociological and economic developments
respectively, had perhaps as much to do with the constitution of visible
nature as an arena of knowledge and pleasure as did the internal development of the sciences of anatomy and natural history. But the new
technology of the microscope had helped open up this arena and had
shown what one could conclude about the relationship of the human

---

[101] See his letter no. 34 to Robert Hooke in Dobell, *Antony van Leeuwenhoek*, pp.
222ff.

[102] Philipp, *Das Werden der Aufklärung*, p. 140.

sensory apparatus to the world, and what followed from this where the old verities were concerned.

Meanwhile, microscopical observation continued to meet with difficulties as it sought to establish itself as a rewarding activity. The romantic stature of microbe discoverers and destroyers in modern times is easy to explain in utilitarian terms. Even so, the charges of triviality and lack of moral purpose are still part of the science critique of the present day, and occasionally scientific practice is still situated in a religious context of worship and mystery; the proportions of devout sincerity and pragmatic cloaking may not have altered as much as one might think. But the substitution of natural wonders for the wonders of revealed religion turned out to have been a sort of gamble that, at the outset, may have helped silence the skeptics, but that, when it failed, left religion weaker than ever, insofar as it had consented to move the field of argument ever farther from its old roots in fideistic conviction and scriptural authority.

Physico-theology is not interesting simply as an interim episode in the secularization process and in the history of the fortunes of religious feeling and apologetics, but as an answer to the question, how does the book of nature come to be a book? How do the *subvisibilia* become the subject of literary and pictorial representation? Moffett's preface to his book of insects shows that the desire for certain forms of pleasure and instruction did not preexist and could not be taken for granted; the creation of the natural history text involved a certain effort at ingratiation. That we no longer insist that science be agreeable or charm us with its moral lesson is proof of the importance we attach—for whatever reason—to fact. But this interest was already in the making in early modern natural history: it is a matter, after all, of talking about what you want to talk about, and trying to create, simultaneously, an attitude of acceptance in the reader.

Kepler, in his elegant treatise on the snowflake, which goes beyond its ostensible subject to consider geometry in nature generally, makes this a matter of high artifice. His New Year's gift to his patron will be from a man who has nothing and will receive nothing to a nobleman who loves nothing. "Whatever it is that attracts you ... must be both exiguous and diminutive, very inexpensive and ephemeral, in fact, almost nothing."[103] He rejects various objects as too big or too small: the sand grains of Archimedes, particles of dust, ten thousand of which are in the heart of every poppy, the invisible pyramids composing the sparks of fire, the acarus mite. Only a happy accident interrupts his musings when

---

[103] Kepler, *Six-Pointed Snowflake*, p. 3.

a snowflake falls on his sleeve: "Here was something smaller than any drop, yet with a pattern." The wit of the story is that Kepler did not make his patron a present of a snowflake but of a treatise on the snowflake. Small things may still have no value, but, from that point forward, a discourse on them does.

7

The Microscope Superfluous and Uncertain

THAT THE INTRODUCTION of the microworld into the subject matter of natural philosophy should have been welcomed by those traditionally designated rationalist philosophers, and regarded with skepticism and mistrust by empiricists, is a surface anomaly whose deeper logic, with all its consequences for the later development of the philosophy of science, needs uncovering. The dialectics of surface and essence, sign and substance, visible alteration and invisible force that furnish so much material to the systematic philosophy of science appear in many forms, in ancient medicine and astronomy, in early modern science, and in contemporary encounters with problematic *subvisibilia*. General questions of knowledge, error, and representation presuppose these concepts and their opposition.

Vasco Ronchi has argued in numerous papers on optics that lenses were not systematically employed before the early seventeenth century because they were not seen as a means of acquiring knowledge, but as tricky devices that distorted appearances and confused the observer. According to Ronchi, when optical instruments were considered by mathematicians and philosophers they must have reasoned as follows: The purpose of sight is to inform us about how things are. But lenses show us objects as larger or smaller, nearer or farther away, than they are in reality; they invert objects, multiply them, and change their colors. Hence if we want to pursue knowledge and avoid deceptions we should stay away from optical devices.[1] On Ronchi's account, medieval philosophers held vision in deep suspicion anyway; they did not understand how visual perception occurs, and they were perturbed by common optical illusions. They regarded the sense of touch as confirming reality, and they understood that touch did not confirm the images delivered by lenses.[2] Artificial vision, Ronchi argues, was validated by Galileo's use of the telescope, and the use of optical instruments in general was validated by Kepler's presentation of a theory of lenses in the *Dioptrics*.

There is undeniable evidence in favor of Ronchi's thesis, as well as much to oppose it. Roger Bacon's interest in optics and his hints about the amazing powers of magnifying lenses in the thirteenth century had

[1] Ronchi, "General Influence of the Development of Optics," p. 125.
[2] Ronchi, "Influence of the Early Development of Optics," p. 197.

resulted, according to Hooke in his *Discourse concerning Telescopes and Microscopes*, in "false Accusations, scandalous Reports, [and] Imprisonment," so that the subject was effectively, in Hooke's words, "quashed" until the time of Porta and Leonard Digges.[3] Henry Cornelius Agrippa had stated meanwhile in his *Vanitie and Uncertaintie of the Artes and Sciences* that glasses were deceits, remembering that Augustine was of the opinion that some "hidden thing" was within them, and he concluded (or pretended to: he then went on to praise Vitelo for his optics) that "all these things are vain and superfluous and invented to no other end, but for pomp and idle pleasure."[4] Francis Quarles moralized about the diversionary aspect of lenses in his devotional poetry,[5] and Galileo's Aristotelian expressed not fear of the demonic, but simply mistrust: "[N]or have I so far put any belief in the newly introduced *occhiale*; indeed, following in the footsteps of other peripatetic philosophers, my colleagues, I have believed to be fallacious or a deception of glasses what others have admired as stupendous operations."[6] In his *Lectures of Light* Hooke himself seemed to state Ronchi's thesis, observing that "it is indeed a great Argument directed against the Use of Telescopes, that there can be no Truth in the Discoveries made by them,"[7] and that many have concluded that all such appearances are uncertain and imperfect, that optical instruments communicate misinformation and create false representations. "[A]ll such objections," he continues, "do only proceed from an Ignorance of the Grounds of Opticks and of Vision, we being equally as certain of the Appearances we discover by them, as of those things which are discovered and seen by the naked Eye."[8]

Ronchi's generalizations have been subjected to vigorous criticism by historians who have pointed out that his claim that the medievals repudiated both natural and lens-assisted sight is doubtful and unsupported. The medievals, it is argued, understood the special conditions that produced illusions and did not think that they invalidated the claims of any sense to general reliability.[9] Eyeglasses, which were invented by 1280, restore and normalize the sight of the aged, a development alleged to have ended the domination of the young in state and church affairs;[10]

---

[3] Hooke, "Discourse concerning Telescopes and Microscopes," in *Philosophical Experiments and Observations*, pp. 257–58.

[4] Agrippa, *Vanitie and Uncertaintie*, p. 37.

[5] Quarles, *Emblems*, book 2, no. 6; book 3, no. 14.

[6] Galilei, *Dialogue concerning the Two Chief World Systems*, in *Opere* 1:366.

[7] Hooke, *Lectures of Light*, in *Posthumous Works*, p. 92.

[8] Ibid., p. 98.

[9] Lindberg and Steneck, "Sense of Vision," pp. 29ff.

[10] "Officials who would earlier have had to retire in their forties or fifties due to failing

medieval philosophers among others would have had to regard them as a boon. The suggestion that Galileo and Kepler as practitioner and theoretician respectively exercised a unidirectional influence on their contemporaries is hardly sustainable. How could Galileo's telescopic investigations have been initiated and accepted, let alone have been the causes of a better reception of the telescope, if sensory observation and instrument-assisted perception were both anathematized by his contemporaries? If the medievals were suspicious of vision in the first place, why would they have contrasted instrument-assisted perception with ordinary perception to the detriment of the former?

The first use of the telescope brought about a period of negotiation in which Galileo and his supporters attempted to defend against his opponents the credibility of his observations of the moon and Jupiter and to justify the novel means by which he had made them. The *Sidereus nuncius* provoked an attack by Martin Horky and a response by the Scotsman Wodderborn and others who argued, with success, that the telescope was not a purveyor of hallucinations.[11] The microscope, by contrast, was not the subject of polemics from its first introduction, and there is no evidence that its theoretical rationalization by Descartes and Kepler was necessary to validate its use as a scientific tool. As Ian Hacking has noted, Kepler did not succeed in giving a physically correct account of the functioning of lenses anyway; that account was supplied long after the microscope had been accepted as a research tool, so the principle that to employ an instrument with confidence one needs a correct theory of the instrument seems to be false.[12] From the historical conjunction of what was, at any rate, an intelligible and useful theory, geometrical optics, and the practical deployment of the lens, one cannot conclude too much about the direction of causal influence.

Even if one could still raise objections about its application to the heavens, the veridicality of the telescope was easy to establish through practical tests such as reading distant inscriptions and counting the windows on faraway buildings. Wodderborn even used the natural credibility of the microscope to defend that of the telescope: "He perfectly distinguished with his telescope [*sic*]," he wrote of Galileo, "the organs of motion and of the senses in the smaller animals. . . . Here hast thou a new proof that the glass concentrating its rays enlarges the object."[13] Art and commerce gave practical reinforcement as well. It is perhaps no accident that the Dutch, who excelled in illusionistic renderings in

---

eyesight could now remain in office for decades, much to the frustration of their juniors." Lindquist, "Wagnerian Theme," p. 160.

[11] Galilei, *Opere* 3:155ff.

[12] See Hacking, "Do We See through a Microscope?"

[13] Galilei, *Opere* 3:164, in Govi, "Compound Microscope," p. 575.

painting and who dominated the cloth trade, were preeminent in scientific microscopy. If a lens enables one to paint a certain texture, to draw an insect, to evaluate the quality of a piece of cloth or a gemstone, one is not tempted to raise theoretical problems about its ultimate reliability. A context of use and acceptance is necessary before such problems can be raised, and they are not solved by convincing proofs or dramatic discoveries, but by practice and correction.

Nevertheless, Ronchi deserves credit for identifying important questions of early modern science: the acceptance of mediated knowledge, and the historical role played by skepticism about lenses. Doubts about the usefulness and reliability of lenses did arise, and they intensified as the instrument gained in power. The picture occasionally presented by Derek J. de Solla Price of a technological innovation clearing away centuries of dispute and conceptual confusion is clearly inadequate: faced with dissonance between inherited beliefs and new experiences, one can always find fault with the experiences, and this in fact happened. Even in the early eighteenth century, optical instruments were still presented in the context of dioptric and catoptric magic. They might be presented now as truth-obscuring, now as truth-revealing, and only gradually did the early modern epistemology of immediate apprehension give way to one of negotiated meaning.

This process was subject to fluctuations: confidence and optimism were followed by extremes of disillusion. As Spinoza said, whoever has loved a thing and commences to hate it, hates it more than if that person had never loved it.[14] Although Bacon regarded the weakness of the senses and the subtlety of nature as hindrances to scientific progress, he did not expect dramatic benefits to follow from sensory enhancement. Hooke accepted Bacon's assessment of the failure of science, but devoted considerable rhetorical energy in 1665 to promoting the microscope as its salvation; by 1692, his confidence would be seriously undermined. In the *Micrographia*, he reasoned that the microscope would directly and simply establish the truth of the mechanical philosophy, a philosophy itself so clear and unambiguous that it was beyond the realm of relativity, misjudgment, and illusion. The microscope would take us down to the realm of philosophical essences; yet this realm of essences would turn out, at bottom, not to be a new world, but a familiar one, the world of machines. With its aid, we would come to discover "the subtilty of the composition of Bodies, the structure of their parts,

---

[14] Spinoza, *Ethics*, part 3, prop. 38. Fournier also argues that "the original incentives for applying the microscope to study nature's smaller structures were . . . defeated by those very investigations." "Fabric of Life," p. 6.

the various texture of their matter, the instruments and manner of their inward motions, and all the other possible appearances of things. . . . we may perhaps be inabled to discern all the secret workings of Nature, almost in the same manner as we do those that are the productions of Art, and are manag'd by Wheels, and Engines, and Springs, that were devised by humane Wit."[15] We would find the productions of nature ultimately as visually and intellectually perspicuous to us in their workings as the products of our own hands.

The occult philosophers knew as well as the mechanical philosophers did that truth lay in the interior of things. But who can see beyond surfaces? Mersenne, in *La veritez des sciences*, had already formulated the paradox.

> Cast your eyes where you will; we see only the surface and the color of things; we taste only their flavor; as for what is underneath it, we do not see a speck: we are like those who are content to touch the garment, to sense the smoke and the shadow; we know nothing of the substance and of the body: and so it was wrong of that dreamer, whom one calls the master of the Peripatetics, to say that the intellect understands the essence of things, since nothing enters the intellect except what has passed in through the senses. And even if we could enter into the interior of things, we would not understand them any better, for we cannot apperceive anything except exterior accidents.[16]

Or, as Fontenelle's philosopher tells his marquise, "All philosophy is based on two things only: curiosity and poor eyesight. . . . The trouble is, we want to know more than we can see." The paradox is that "true philosophers spend a lifetime not believing what they do see, and theorizing on what they don't see."[17] What we see is always an appearance produced by an underlying mechanism, comparable to the stage setting and stage acrobatics produced by wheels, weights, and wires. We are like the observers in the audience who can only speculate—by means of different phenomenon-saving schemes—how the appearances are produced. The philosopher's rule has to be something like this: disbelieve what you see—it is only appearance and is produced by what you do not see. The hope is that this disbelief does not continue forever, that we reach a realm where what we see is simply what is: springs, engines, and wheels, in Hooke's formulation. But if anything at all can be opened up and examined with a microscope, why should we stop our investigation with the springs and wheels of the microworld? The paradox here is similar to one formulated or reformulated by David Hume. In Hume's version, we believe that it is from an inner mechanism, a

[15] Hooke, *Micrographia*, preface.
[16] Mersenne, *La veritez des sciences*, p. 9.
[17] Fontenelle, *Conversations*, p. 12.

hidden arrangement of parts, that the characteristic properties and behavior of a substance flow. But we can never observe this mechanism in
the act of producing its effects. What we see is always only a product;
we cannot perceive production, but must always refer it to some further
invisible mechanism that we do not see. Inner powers are always hypothesized and projected; they cannot be rationally apprehended. To
the extent that our science—the paradigm of rational thought—is explanatory, it is irrational.[18]

Was there any association, one wonders, between the theoretical
skepticism that rests on the philosophical distinction between appearances and reality and a skepticism directly linked to the practical problems of achieving knowledge of internal structures and processes? Is
there not a transcendental aspect to the problem of appearance and
reality, powers and explanation, and, unrelated to it, a mundane one?
The assertion of the unknowability of substance or individual essence
with which Locke, Berkeley, and Hume concern themselves has both a
transcendental and an empirical dimension. Hume's paradox is not susceptible of dissolution except through a transcendental theory of the
harmony of our projective faculties with the world. But the possibility
of science does not after all depend on such a theory; where skepticism
about knowledge of nature was motivated by practical and conceptual
difficulties in microscience, its actual purchase on the philosophy of science was far stronger and its interest for the historian greater.

There were a number of obstacles and impediments of a physical and
psychological nature to microscience that were acutely experienced in
its first period. Optical theory shows why there are unavoidable trade-
offs in the construction of any lens. Optimality, not perfection, is all
that can reasonably be sought, and though their deficiencies have been
exaggerated, early optical instruments were less convenient and reliable
than later ones. Naturalists had to find the time and money to engage in
microscopical research and to locate in themselves the resources to construct whatever professional and moral rationalization for the activity
seemed necessary. On the level of observation, there were difficulties
involved in attaining a stable image of the desired specimen and in being able to describe in words or to put down in a drawing what was
seen. On the level of dissemination, there was the difficulty of gaining
access to publication and establishing oneself as a credible and reliable
observer. A full diachronic study of the development of scientific microscopy would have to show how each set of problems was solved
through a series of technical, psychological, and sociological responses,
rationalizations, and innovations.

[18] Hume, *Enquiry concerning Human Understanding*, sect. 4, parts 2–3.

Whereas ambiguity and inconsistency in visual data are largely eliminated by the processing devices that govern ordinary stereoscopic vision, microscopical experience is another matter. As Pieter Harting noted, perceived size depends on the state of adaptation of the eye and varies from observer to observer. Most microscopical images are multiply readable. Thin stripes can be seen as indentations, ribs, boundaries of fibers, boundaries of hollow tubes, or the edges of scales; a ring can be a slice, a drip, a bubble, an opening, or a thick or thin place on a membrane.[19] "I beg your favour, Sir," Leeuwenhoek wrote to the secretary of the Royal Society, "to communicate this to Dr. *Grew*, with my service to him, and to inquire of him, whether he hath seen as well as I, whether the great Vessels or pores, that are exprest by him in his figures, do not consist of globuls; . . . as also that in the same [figure] do lye oblique membranes or films, by me call'd valves . . .; again whether the particles of the Wood be not all of them very small Vessels or pores."[20] Swammerdam cautioned that "those who are fond of the microscope must take care not to confide in one lens or glass only, and must not always view the object in the same manner and situation, for by this means many errors arise."[21] And Hooke noted in the *Micrographia* that there is a persistent problem of determination in microscopical work:

It is exceedingly difficult in some Objects, to distinguish between a *prominency* and a *depression*, between a *shadow* and a *black stain*, or a *reflection* and a *whiteness in the color*. Besides, the transparency of most Objects renders them yet much more difficult than if they were *opacous*. The Eyes of a Fly in one kind of light appears almost like a Lattice, drill'd through with abundance of small holes. . . . In the Sunshine they look like a Surface cover'd with golden Nails, in another posture, like a Surface cover'd with Pyramids; in another with Cones, and in other postures of quite other shapes.[22]

Observation is fraught with mundane perils: scratches on the glass, impurities and irregularities in it, dust, "entoptical appearances" from the eye itself. Strong magnification makes color, hence structure, seem to disappear: to be made visible, parts must be artificially stiffened and colored. But in this case one cannot claim so simply to be seeing nature rather than an artificial preparation derived from a natural thing. Our teeth appear white or yellow to the naked eye, but sections of them look brown or green under the microscope. To describe the appearance of something under the microscope is, Berkeley decided, to describe a mi-

[19] Harting, *Das Mikroscop* 1:329.
[20] Leeuwenhoek, "Concerning the Texture of Trees," p. 656.
[21] Swammerdam, *Book of Nature*, 1:157.
[22] Hooke, *Micrographia*, preface.

croscopical appearance, not to give a more accurate, better description of it.[23] Special problems are presented by living specimens. Anyone who has tried to reproduce Leeuwenhoek's pepper water experiments, even with an easily focused modern student microscope, appreciates at once that the speed with which infusoria move across the field at high magnifications makes it difficult to observe them: by slowing them down you lose their structure, and by killing or incapacitating them you lose the chance to study their motion and behavior and they begin to disintegrate.

Or else organisms may appear where there are none: the worms reportedly found by so many observers in parenchymatous tissue, historians who have taken the trouble to experiment with antique, non-achromaticized microscopes have suggested, are optical artifacts.[24] The appearances giving rise to the notorious nineteenth-century "fibro-reticular" theory of tissue structure, as well as such misleading oddities as hairs in dental enamel and glandular follicles on the surface of the brain's cortex, were further examples of what Luigi Belloni and Bruno Zanobio call "illusory micrography."[25] Verbal descriptions, meanwhile, might overshoot the mark in an attempt to recreate an interesting visual experience or please the reader, or undershoot the mark, or both. It is difficult to say whether descriptions such as this one by Henry Power succeeded in communicating any definite visual impression to the reader at all: "The eye of a Bee is of a protuberant oval figure, black and all foraminulous, drill'd full of innumerable holes like a Grater or Thimble; and, which is more wonderful, we could plainly see, that the holes were all of a square figure like an honey comb, and stuck full of small hairs (like the pores in our skin). . . . Now these holes were not absolute perforations, but onely dimples in their crustaceous *Tunica Cornea*; which it seems is full of little pit holes, like the cap of a thimble."[26] Although illustrations reduce indeterminacy, they do not eliminate it, for the illustrator must either reproduce the ambiguous feature ambiguously or guess at the correct alternative. If the verbal description tends to be

[23] Berkeley, *Dialogues*, in *Works of George Berkeley* 2:245.

[24] Belloni, "Athanasius Kircher."

[25] Zanobio, "L'immagine filamentoso-reticolare"; Belloni, "Micrografia illusoria e 'animalcula.'" In fact, concentration on the outer wall or membrane of the cell, rather than its contents, was understandable in those who were looking for the actual building blocks of creatures. The fibro-reticular theory was encouraged not simply by indistinct images, but by the common metaphor of nature's weaving, which we find in the modern term "tissue" and before that in Aristotle's reference to the embryo as like the plaited net of Orpheus; in Hooke's description of the "looms" of nature on which its forms are woven; and in Grew's description of the plant stem in cross-section as like lace woven by women.

[26] Power, *Experimental Philosophy*, pp. 3–4.

poeticized, and the image deflected toward a known object of compari-son, the drawing too invites distortion in the direction of the known. Gerald Turner notes that, before the invention of photomicrography in 1839, each available means of producing a visual image of what was seen was unsatisfactory in its own way. One could draw a remembered image, moving back and forth between instrument and paper, or at-tempt to keep one eye on the paper next to the microscope, one eye on the microscopical image, or employ a projection microscope and trace the image.[27] Photographs, by contrast, fuse the processes of reception and image creation; they are quickly made, easy to reproduce, and in-sinuate fewer characteristics of personal style into the representation. Still, drawings can show detail that photography cannot, for the focus can be varied by the human observer but not by the camera to obtain a better representation of three-dimensional structure.[28]

The new, open science that was the creation of the academies en-forced an idea of accountability. The academy set itself up as an arbiter able to discredit particular observers whose reports did not hold up. At the same time, by being able to confer prestige on and lend support to talented discoverers, it put temptation in the way. Kircher was an exam-ple of an observer who achieved celebrity as a virtuoso and notoriety as a fantasist with amazing experiments and observations; his association with the theory of animate contagion probably did that theory no good. Yet, like the extravagant Kenelm Digby, he was never excluded from the community of protoscience; his books were reviewed, his opinions cited. Outright fraud of the sort seen in the celebrated case of M.A.C.D., and parody, exemplified in the case of the famous de la Plantade illus-trations of the baby-enclosing spermatozoon, was another matter. Leeuwenhoek had known a charlatan who claimed to have a micro-scope so excellent that one could see in it "not only mites . . . but the atoms of Epicurus, the subtle matter of Descartes, the vapours of the earth, those which our bodies transpire, and the influence of the stars;" a coat viewed through this apparatus seemed to be in the process of being devoured by worms. Leeuwenhoek exposed this fraud, and showed that holes had been ground in the lenses into which minute objects could be placed, which would be seen projected against the background.[29] More common than fraud were examples of "oversee-ing" or "overrepresenting." Though cases of seriously misleading re-portage were infrequent when compared to the number of careful and

[27] Turner, "Microscopical Communication," p. 4.
[28] Ibid., p. 9.
[29] Leeuwenhoek, letter 139, *Arcana natura*, in Miall, *Early Naturalists*, p. 222.

unsensational drawings and descriptions, their occurrence in association with issues of great existential interest, generation and contagion, gave them additional resonance.

Did the numerous corrections of overseeing and overrepresenting that plant and animal anatomists routinely made of one another's work imply objectivity and the possibility of accurate observation and recording? Or were they cause for skepticism about the whole enterprise? There is a fine line between a cooperative endeavor that advances through mutual correction on minute points of detail and an empty scholasticism in which such disputation reflects the lack of any determinate subject matter. The more exotic claims of early microscopy—such as Croone's claim to have seen a whole chicken lying underneath a fold of membrane in an unincubated egg, and Hartsoeker's suggestion that the fetus might be enclosed in the spermatozoon—were not announced dogmatically, but with considerable reservation. Croone admitted that some people might "consider the trustworthiness of these observations as extremely slight or conclude that I had been dreaming."[30] Nicolaas Hartsoeker's drawing of the curled-up baby was intended to represent, as he stated clearly, not what he had seen but what one might hope to see.[31] But when the language of the microscopists themselves was so hesitant and conjectural, and when their disagreements were nevertheless so pronounced, what outsider would not have been tempted to believe that natural philosophy was a thoroughly doubtful enterprise? Taken narrowly, it certainly was, as insiders such as Leeuwenhoek readily admitted:

> No Body must Publish or bring to light, new Discoveries, and judge by one sight, but he must see the same over and over several times, for it doth happen often to me, that People looking through a Magnifying-glass, do say now I see this, and then that, and when I give them better Instructions, they saw themselves mistaken in their opinion, and what is more, even he that is very well used to looking through Magnifying-glasses may be misled by giving too sudden a Judgment, of what he doth see.[32]

Leeuwenhoek's first communication to the Royal Society concerned what he regarded as an error in Hooke's description of the bee in *Micrographia*. Soon he himself was being challenged. Writing to his father about Leeuwenhoek's observations on blood, milk, sweat, fat, and tears, which were published in the *Philosophical Transactions* for 1674, Christiaan Huygens complained, "I have tried in vain to see certain

[30] Croone, "De formatione pulli in ovo," in Cole, "Dr. William Croone on Generation," p. 134.

[31] Hartsoeker, *Essai de dioptrique*, pp. 229–30.

[32] Leeuwenhoek, "Concerning the Animalcula in *Semine humano*," *Philosophical Transactions* 21 (1699): 306.

things which he sees, but I doubt they are more than optical illusions. . . . he pretends to discover the particles of which water, wine and
other liquids are composed."[33] When Leeuwenhoek's microscopes were
studied after his death, Henry Baker found himself reluctantly forced to
conclude that Leeuwenhoek had either lied about his observations or
declined to present the Royal Society with his best instruments. "I am
fully persuaded," Baker says, "and believe I shall be able to prove, that
many of the Discoveries that Leeuwenhoek gives an Account of, could
not possibly be made by glasses that magnify no more than this." With
a lens magnifying two hundred diameters, an object a million times as
small as a grain of sand would appear too small to enable anyone to
delineate the particles clearly. This, Baker thought, rendered Leeuwenhoek's claim that, if the tails of his spermatic animalcula had been
forked, he would have seen it, doubtful, and it certainly exploded his
claim to have seen vessels in the eye 18,399,740 times as small as a
grain of sand. Faced with an unpleasant choice of alternatives, Baker
decided that Leeuwenhoek had after all been honest in his reports but
had sent inferior microscopes, display pieces only, to the society.[34]
Lorenzo Bellini, a colleague of Malpighi, was less charitable; he did not
care for Leeuwenhoek as an observer at all. "While I have nothing to
say about these hearts and eyes of his and would concede to him everything he wants, I do not see that these worms in the semen have been
well proved, much less the way food is digested. . . . I wonder whether
this man has not assisted his eyes and his microscope with some prepos-
sessions that have made him see things wrongly. Yet it may all be so,
and I shall await proof of it."[35] But these doubts and criticisms within
the microscopical community itself did not seriously challenge the assumption that the microscope was a truth-revealing instrument. Even
Kerckring's warning about microscopes in his *Spicilegium anatomicum*
of 1670 was not meant to be as discouraging as his memorable chapter
title, "Per microscopium incertum in anatomia judicium," suggests. He
is troubled not by the inadequacies of the instrument but because the
marvels and wonders that can be seen in the structures of tissues can
never be fathomed by the human mind, which is unable to comprehend
the work of God in its smallest details.[36]

The momentum of the 1670s and 1680s ebbed rapidly, and by 1692
Robert Hooke had painted a surprisingly dismal picture of the state of

[33] Huygens, letter to Oldenburg, no. 2003, 30 January 1675, in *Oeuvres complètes*
7:399.

[34] Baker, "Account of Mr. Leeuwenhoek's Microscopes," p. 507.

[35] Bellini, letter to Malpighi, 8 October 1686, in Adelmann, *Marcello Malpighi* 1:506.

[36] Kerckring, *Spicilegium anatomicum*, pp. 178–79.

applied optics in his *Discourse*, suggesting that it was a fashion whose time had come and gone. "Tho there has been some Life left in the Grinders of Glasses," he says, referring to telescopy,

> yet the Warmth of those, that should have used them, has grown cool; and little of new Discoveries hath been made by them. . . .
>
> Much the same has been the Fate of Microscopes, as to their Invention, Improvements, Use, Neglect, and Sighting, which are now reduced almost to a single Votary, which is Mr. *Leeuwenhoek*; besides whom, I hear of none that make any other Use of that Instrument, but for Diversion and Pastime, and that by reason it is become a portable Instrument, and easy to be carried in one's Pocket.[37]

The feeling of the present generation is, he says, that the "Subjects to be enquired into are exhausted, and no more is to be done," that there is nothing profitable, able to "bring ready Money," in microscopical researches. This Hooke found regrettable. "As to the discoveries that may be made," he says, in both telescopy and microscopy, "I conceive they are vastly greater, both for Number and Value, than those few that have been already made; and not only for the Information of the Intellect, but what answers their greatest Objection, even for the increasing their Treasure."[38]

Still, the scientific community had evidently voted with its feet, and there was nothing in Hooke's *Discourse* to excite anyone about taking up the instrument again. The history of microscopy that he gives is selective and oddly distorted. The references are out of date: Francesco Stelluti's bee of 1625 is prominently mentioned, as are Giovanni Hodierna's observations from the 1640s and those of an obscure Panarolla[39] from the 1650s, but Hooke does not refer to his own work, to Malpighi's, or to that of the prominent Dutch anatomists—Swammerdam, van Horne, Steno, Ruysch, de Graaf, or the younger Bartholin—noting only that "Many others were also found to make some few Observations in other Countries; but by Degrees, it is become almost out of Use and Repute."[40]

This cooling off on Hooke's part, his indifference to the excitement and controversies of the previous two decades, his pushing microscopy back into the more remote past, calls for explanation. His perception of the situation was not inaccurate: the valuable study by Marian Fournier

[37] Hooke, "Discourse concerning Telescopes and Microscopes," in *Philosophical Experiments and Observations*, p. 261.

[38] Ibid., p. 262.

[39] Identified by M. Fournier as Domenico Panarolo, who referred to microscopical discoveries in his medical text of 1652; "Fabric of Life," p. 48.

[40] Ibid., p. 268.

shows that the publication of papers on microscopy reached a peak in the early 1680s, declining steeply thereafter.[41] Leeuwenhoek was indeed the major contributor to the literature after 1690, and was responsible for the vast majority of publications issued between 1715 and 1725. In 1691, Swammerdam was dead; Malpighi published his last work in 1689 and would die in 1694; Grew published nothing new on plants after 1684, turning to cosmology and theology. Where Hooke's own short career as a microscopist is concerned, we must remember that microscopy was, after all, only one subject among many, and that the Royal Society itself was flagging as an institution before Newton assumed its presidency in 1703 and refocused its interests toward physical science—mechanics, heat, light, water, air, electricity, magnetism, and meteorology.[42] Illustrations of microscopic objects were still competing meanwhile with other, more-dramatic images in the pages of proto-scientific academy publications. Grew published his cross-sections of plant stems in the German *Miscellanea curiosa*, and these regular geometric figures, though not without visual interest, had to share space with a monstrous chrysanthemum, a monstrous lamb, a vegetable shaped like a pair of praying hands, a two-headed calf, a bearded woman, voided stones, and other natural and artificial wonders.[43]

As Hooke said, the microscope became popular, a toy for diversion and recreation wielded by observers with no particular interest either in enlarging the field of microscopical research or in settling controversies with its help. Henry Baker's *Microscope Made Easy* of 1743 reflects Hooke's feeling that all the important microscopical discoveries have already been made and suggests that the best an author can do is to assist the home experimenter in reproducing them. The awkwardness of managing for oneself in this regard was overcome as manufacturers increasingly did the work of choice and preparation in advance. Through the eighteenth and nineteenth centuries, microscopes were sold with prepared "sliders," which furnished a preselected range of specimens mounted between mica or (later) glass slides. But, as with any packaged mix, boredom and a dulling of curiosity now replaced incompetence, as Baker realized:

Many, even of those who have purchas'd Microscopes, are so little acquainted with their general and extensive Usefulness, and so much at a Loss for Objects to examine by them; that after diverting themselves, and their Friends, some few times, with what they find in the Sliders bought with them,

---

[41] Ibid., p. 187. Cf. the graph on p. 16 of her study and her conclusions in chap. 8.

[42] Heilbron, *Physics at the Royal Society*, pp. 1–2.

[43] Such publications include the Amsterdam *Collectanea Medico-Physica* and the *Miscellanaea Curiosa sive Ephemeridum Medico-physicarum Germanicarum Academiae Naturae Curiosum.*

or two or three more common Things, the Microscopes are laid aside as of little further value; and a Supposition that this must be the Case prevents many others from buying them.[44]

This was in Baker's view lamentable: no other invention in the history of the world, he thought, could perhaps be found "so constantly capable of entertaining, improving, and satisfying the Mind of Man." Such urgings probably did not help. The feminization of the microscope also assisted in reducing the prestige of an instrument of interest to and usable by ladies.[45] And satire had its effect. Samuel Butler's *Hudibras*, written "in the time of the late wars," and *The Elephant in the Moon* made fun of natural historians and their preoccupations,[46] and the hero of Thomas Shadwell's *Virtuoso* (1676), Sir Nicholas Gimcrack, is described by his niece as "one who has broken his brains about the nature of maggots, who has studied these twenty years to find out the several sorts of spiders, and never cares for understanding mankind."[47] The virtuoso's motto is "so it be knowledge, 'tis no matter of what,"[48] a line that gives pause even today. Sir Nicholas discovers that the blue of plums consists of microscopical living creatures; this blue, like Hooke's program for science, "comes first to fluidity, then to orbiculation, then fixation ... then crystallization, from thence to germination or ebullition ... animation, sensation, local motion, and the like."[49]

In the first flush of enthusiasm for the microscope, too much had been hoped for too soon. According to Abraham Cowley's *Ode to the Royal Society*, published in Sprat's *History of the Royal Society* (1667), nature had inscribed its secrets in fine print: "You've learned to read her smallest Hand / And well begun her deepest Sense to understand."[50] But neither statement rang true: neither the reading nor the understanding met expectations. The systematic science of form that Hooke and Power had imagined in the mid-1660s looked no more attainable. Hooke thought that magnification would result in simplification, that it would reveal elements or fundamentals, so he attempted in the *Micrographia* to transform microscopical observation into something else—into geometry or even grammar. In the preface he suggested that the microscope would produce an "alphabet" for the expression of complex forms. "As in geometry we begin with bodies of the most simple nature, so we need begin with *letters* before we try to write *sentences* or draw pictures."

[44] Baker, *Microscope Made Easy*, p. 51.
[45] See Nicolson, *Science and Imagination*, p. 190.
[46] Ibid., p. 171.
[47] Shadwell, *Virtuoso*, 1.2.11–13.
[48] Ibid., 3.3.26–27.
[49] Ibid., 4.3.221–28.
[50] Sprat, *History of the Royal Society*, preface.

Beginning with "Fluidity, Orbiculation, Fixation, and Angularization, or Crystallization," we may proceed all the way to "Germination . . . Vegetation, Plant Animation, Animation, Sensation and finally . . . Imagination." From the forms of crystals, we may move to those of mushrooms, thence to plants in general, and so on, "these several inquiries having no lesser dependence on one another than any select number of propositions in Mathematical Elements may be made to have."[51]

Against the view that knowledge of nature could be acquired through the process of visual analysis and synthesis, Joseph Glanvill had argued and continued to argue, his approval of Hooke and microscopy notwithstanding, that the alphabet of nature, the alphabet of simple forms, cannot be read. We can reason only by analogy to what we experience on a larger scale, but the *difference* of the microworld makes these results inadequate and misleading:

> 'tis no doubt with the considerate, but that the *rudiments* of Nature are very unlike the grosser *appearances*. Thus in things obvious, there's but little resemblance between the *Mucous sperm*, and the compleated *Animal*. The *Egge* is not like the oviparous production: nor the corrupted much like the *creature* that creeps from it . . . nor do *vegetable derivatives* ordinarily resemble their *simple* seminalities. So then, since there's so much dissimilitude between *Cause* and *Effect* in the more palpable *Phenomena*, we can expect no less between them, and their *invisible* efficients.[52]

Boyle, too, despaired of understanding surface effects with the help of an underlying alphabet of forms and interactions and retreated to argument by analogy; and Locke, in his turn, offered his endorsement of analogy, but in a way undermined by his misgivings about the whole enterprise of looking for real essences.

What did these writers have in mind by the term "analogy" in this connection? In corpuscularian texts, macroscopic experiences are invoked in order to give the imagination a clear picture of an imputed microprocess: the air was said to be composed of a multitude of tiny springs; acid particles were said to be sharp, water particles, flexible, and so on. Boyle frequently invoked macroscopically appreciable alterations in qualities that result from changes in motion or situation in order to model microscopic alterations: velvet, waving wheat, the hair of angry dogs, the pounding of glass and scraping of horn, pea fields viewed from the side, the nap of taffeta, all show us the origin of forms and qualities.[53]

---

[51] Hooke, *Micrographia*, preface.

[52] Glanvill, *Vanity of Dogmatizing*, p. 211.

[53] Boyle, *Experiments and Considerations Touching Colours*, in *Works* (1744 ed.), 5:678–79.

As we suppose that macroscopically visible surface textures change the appearance of silk and velvet, Locke says, so we may suppose that "the Colour and shining of Bodies, is in them nothing but the different Arrangement and Refraction of their minute and insensible parts." A conventional statement, so far. But a good picture of Locke's alienation from microscopical science is given by his suggestions that subvisible corpuscles and supernatural beings pose similar problems for epistemology and must be addressed by similar means, by analogical reasoning. For he goes on in the same paragraph to argue that, as there are creatures beneath us in the scale of nature, there must also be rational beings above us. "This sort of Probability, which is the best conduct of rational Experiments, and the rise of Hypothesis, has also its Use and Influence; and a wary Reasoning from Analogy leads us often into the discovery of Truths, and useful Productions, which would otherwise lie concealed."[54]

This promotion of analogy as a means to knowledge of nature has a strange unsatisfactoriness, deriving on one side from the juxtaposition of nap-phenomena and angels, on the other side from the vague and here unsubstantiated claim that analogies have led to truths and productions. There is a failure to address Glanvill's point that analogies are useless in the solution of the truly interesting and difficult questions. And, in fact, analogy is a treacherous resort for a program committed to reductionism, and scornful of the tautologous nature of explanations in terms of forms and qualities. Analogies invite a regress, as in the notorious case of the "spring of air," or when Malebranche argues that old people learn less readily because of the stiffness of the fibers in their brains. One feels in the latter case that the method of analogy is simply phenomenology: though the mind may feel "stiff" at times, only a further account stated in terms of microprocesses could explain why stiff fibers interfere with learning rather than facilitating it. But the real problem is that while analogy might suggest hypotheses—and faith in the microscope meant that one was prepared to say that reality could only be constituted in one way, only one hypothesis could be true—any direct confirmation of these hypotheses seemed impossible.

What faced the microscopist, except in those lucky instances in Leeuwenhoek's and Malpighi's science in which structure and function seemed to explain one another, was the absolute unintelligibility of the structures perceived: their scientific muteness. Microscopy was no replacement for the scholastic vegetative, motive, and nutritive faculties; it could show muscle fibers and vessels for sap, but it did not really show how much of anything could happen. Grew, who filled pages with

---

[54] Locke, *Essay concerning Human Understanding* 4.16.12.

detailed drawings of plant parts as seen through a microscope, was forced to ask himself the question, what is this structure for?

> For when upon the Dissection of *Vegetables*, we see so great a difference in them, that not only their Outward *Figures*, but also their inward *Structure*, is so Elegant; and in all so Various; it must needs lead us to Think, That the Inward *Varieties*, were either to no *End*; or if they were, we must assign to what. To imagine the first, were exceeding vain; as if Nature, the Handmaid of Divine Wisdom, should with Her fine Needle and Thread stitch up so many several Pieces, of so difficult, and yet so groundless a Work. But if for some End, then either only to be looked upon or some other besides.[55]

But to be looked upon could not be their purpose, Grew said, because some of their tiniest parts would remain forever unseen. Hence we must suppose, without really understanding how this can be, that their purpose really is vegetation. It is not clear at this point where to turn. Greater magnification, a sharper image, more convenient microscopes do not seem to be the answer: more images amount to more mystification. The microscopist who examines a plate with lenses of successively higher power sees new structures emerge and the earlier ones disappear from view. The suggestion that what appears with more magnification might explain what happens at a level closer to the surface is troublesome, for to get one level in view is to lose the other.

So the microscope again seems to offer only a series of unconnected views. The fading of significance at the microlevel is confirmed by the draining out of the qualities of the image. As Boyle had pointed out, the microscope does not intensify the colors of ordinary experience but fades them. Opaque bodies grow transparent; blood grows gray. The latent image—supposedly nearer the real essence—is fading, on Boyle's account, out of sight altogether: "[T]he leaves will afford the most transparent sort of consistent bodies ... and a single leaf or plate will be so far from being opacous, that it will scarce be so much as visible. And multitudes of bodies there are, whose fragments seem opacous to the naked eye, which yet, when I have included them in good microscopes, appeared transparent. ... I am not yet sure that there are no bodies, whose minute particles even in such a microscope as that of mine, will not appear diaphanous."[56] Power saw that particles of ashes and sand viewed through the microscope seemed to be made of transparent glass, "not dirty, opaque and contemptible," and Newton con-

---

[55] Grew, *Anatomy of Plants*, p. 8.
[56] Boyle, *Experiments and Considerations Touching Colours*, in *Works* (1744 ed.), 5:690.

firmed that "the least parts of all bodies are in some measure transparent."[57]

Meanwhile in Italy the medical professors at Bologna were mounting an attack on Malpighi's microanatomy. In 1689, Paolo Mini, a former student of Malpighi's turned renegade, published four theses that were publicly disputed on 13 January. The first conceded that God had prepared a noble, inspection-worthy domicile for the soul in the human body but stated that "it is our firm opinion that the anatomy of the exceedingly small, internal conformation of the viscera, which has been extolled in these very times, is of use to no physician."[58] The second thesis stated that Borelli's fundamental iatromechanical principle, that the humors are separated in the body "solely by means of a structure that behaves like a sieve," was absolutely untrue. The third thesis was directed against comparative researches on insects and plants: the understanding, "arrived at by the exquisite resolution of the parts composing them, is certainly an outstanding labor of our times. But whatever or how great the labor, it is the store of philosophy, not of medicine that it enriches. For a knowledge of the marvelous conformation of these entities will not advance the art of curing the sick." The recommendation afforded to the young physician was rather to "learn the difference between diagnostic and prognostic signs and symptoms and the positions of their organic parts, by means of which the names of diseases and their periods and outcomes are known. For this is the only anatomy of the human body that is useful" (ibid.).

Mini's theses formalized and intensified charges Malpighi had been hearing for years, and he found them troubling. "I cannot deny," he wrote, "that my spirit which for more than thirty-five years has been incessantly and always intolerantly harassed by those emulous of me, has been enervated by these last maneuvers. . . . I foresee no end to these troubles for the hotheadedness, madness and hatred are passing from the masters to the scholars, whose audacity is growing because of youth and because they have nothing to lose. . . . scholarly affairs are deteriorating to such an extent that we are resting against a leaning wall" (pp. 534–35). In May 1689, a book entitled *De recentiorum medicorum studio dissertatio epistolaris ad amicum* appeared, written by another old opponent, Girolamo Sbaraglia, although published anonymously. In the work, as Malpighi summarized it, "the author, assuming the empirical sect [to be the correct one], attempts to damn

---

[57] Power, *Experimental Philosophy*, obs. 33; Newton, *Opticks*, book 2, part 3, prop. 2, p. 248.

[58] Adelmann, *Marcello Malpighi* 1:533ff. Subsequent references to this work are cited parenthetically in the text.

anatomical exercises carried out with the microscope *ad minima* on man, animals, and plants, and censures and ridicules them as useless and as abortuses of the mind's unlawful desires" (p. 556).

Sbaraglia tells the story of Alexander the Great, who gave a measure of chick peas as a prize to a soldier for winning a game involving impaling chick peas on the point of a needle; idle diversions are permissible, he says, in case one feels like doing nothing else, but there is nothing praiseworthy about them (p. 564). The points developed in the treatise are as follows: fine anatomy is useless in warding off disease; in fact, it does not answer basic questions about the use of parts. It is important to understand the functions of the various organs—kidneys, breasts, testes, and so forth—but, once these have been established, further investigation of their structure teaches us nothing new. We still do not understand the function of the pancreas and the spleen, and a study of their structure has not been enlightening on this score. We know what the brain looks like but nothing about how it functions: "the more it is subjected to the anatomist's knife, the more obscure its use becomes," and the lungs are still mysterious; we know that animals breathe and need to breathe, but we do not know why (p. 561). We have no better cures for diseases of the brain and lung than were available in ancient times; and pleurisy, asthma, and pneumonia are still intractable. Medications that are commonly prescribed have been discovered accidentally and are administered "because of similitude: the finer structure of the viscera has certainly not pointed them out." Nature is able to heal itself in the absence of our theoretical understanding, and the physician's art may be blind. "Just as broken, luxated, and corrupted bones are made well by skill, though the structure of bone is not known, so, too, the other parts are probably restored." Good medicines, moreover, are generally efficacious throughout the body, "and if a particular remedy is required, use, experience, and analogy will suggest it; it will not be deduced from the finest component of the parts" (ibid.).

Comparative anatomy, Sbaraglia added, is also useless to medicine. Studies of the internal organs of women and the females of other species have produced no remedies for reproductive disorders. "The indications deduced from them teach us merely new verbal connections between the indications and the observations, but these connections exist only in the mind, not in Nature" (p. 562). Plant anatomy, and especially the activity of comparing parts of plants to animals, teaches us nothing of the specific virtues of plants. "I have no hesitation in thinking that the dignity of things is lowered. . . . These studies seem to me to be gardens of Tantalus, where things which appear to be something are actually nothing, so that they detain our minds by a kind of sweetness; but after the damage has been done, it is recognized that for medical practice all these things are inanities" (pp. 562–63). The microanatomist has got lost in what after all is only a kind of surface structure:

> The nettle does not owe to the microscope its faculty of restraining the force of the blood, though this instrument reveals its prickly surface. The schools do not profit from the virtue of the seed of the white poppy because the eye armed with the microscope has seen that its surface is provided with many square, pentagonal, and hexagonal figures; the seeds of henbane, portulaca, and thyme are not of medicinal value because the microscope has revealed their use. . . . Everyone knows that no solid advantage to medicine has proceeded from such studies. (p. 563)

Sbaraglia concludes by predicting oblivion for Malpighi and his followers.

The last years of Malpighi's life were taken up with composing a reply to this document. He aired his grievances in the beginning, citing the number of works that had been written against him and the number of lectures attacking him. He responds to the charge of vaulting ambition by pointing out that his work was commanded by the Royal Society and that everything he did "was done to escape the tedium of a life afflicted by poor health." To defend himself he must turn back to philosophy, and to Borelli's conception of the body. Nature, to carry out its marvelous operations, has constructed plant and animal bodies "with a very large number of machines, which are of necessity made up of extremely minute parts so shaped and situated as to form a marvellous organ, the structure and composition of which are usually invisible to the naked eye without the aid of the microscope. . . . Just as Nature deserves praise and admiration for making machines so small, so too the physician who observes them to the best of his ability is worthy of praise, not blame, for he must also correct and repair these machines as well as he can every time they get out of order" (p. 568). The machines of our body, which good medicine must understand, are made up of cords, filaments, beams, levers, tissues, fluids coursing here and there, cisterns, canals, filters, sieves, and similar mechanisms. By studying these, anatomists have succeeded in forming models of them, by means of which they demonstrate the causes of effects and give the reasons for them a priori, and so they progress to physiology, pathology, and medicine.[59] Borrowing a figure ostensibly of Robert Boyle's, Malpighi argues that if the soul acts on the body in growth, sensation, and motion, it is "forced to act in conformity with the machine on which it is acting, just as a clock or a mill is moved in the same way by a pendulum of lead or

---

[59] "Riposta del Dottor Marcello Malpighi alla lettera intitolata *De recentiorum medicorum studio dissertatio epistolaris,*" in Adelmann, *Marcello Malpighi* 1:570.

stone, or by an animal, or a man; indeed if an angel moved it, he would produce the same motion with changes of position as the animals or other agents do" (p. 571). If the machine is out of order, one should try to repair the wheels or damaged structure without bothering about how the angel would approach the problem (ibid.). We cannot cure a damaged faculty of the soul, but we can remove the forces that destroy or interfere with the operations of machines.

Against those who claim that theoretical medicine has produced no new treatments or remedies, Malpighi presents counterexamples. Catarrhs are better understood through the knowledge of the glands of the mouth and throat, he argues, and asthma and phthisis are better understood in light of appreciation of the membranous structure of the glands; knowledge of the accumulation of fluid in the "glandular pericardium" has also helped people comprehend this ailment. Treatments in kidney disease have changed as a result of the appreciation of the kidney's fine anatomy; plenty of water rather than astringents and refrigerants is the new recommendation. The frightening oily substance on top of urine that seems to be the very substance of the patient's body has been shown by the microscope to be a collection of sand particles, with a reassuringly less drastic treatment called for (p. 577). Although brain maladies are no more treatable, it is now known that they result from the misdirection of juice produced by the cortical glands; the mechanisms of sight, hearing, smell, taste, and touch have all been clarified. There are too many medicines anyway, most of them dangerous; yet Sbaraglia wants more and can only get them from empirics, who get them by irrational means, from signatures (p. 579). Malpighi admits, however, that healing is no longer the first concern of the modern student of anatomy. The microanatomists are "gripped by curiosity, sickened by the tedium of practice, and appalled by the vastness of the art of medicine." The leisure and freedom they require in turn does not encourage their clinical practice. This is a major concession to Sbaraglia. Malpighi concedes that microanatomists are pursuing philosophy and not medicine; they can argue only with some difficulty for their inclusion within the disciplinary boundaries of medicine and the institutional boundaries of the university medical faculty.

The comparison between plants and animals, Malpighi says, addressing the attack on comparative anatomy, does not detract from the dignity of the latter. It is not necessary to believe that plants have sensation and motion to see that, in their growth and nutrition, they resemble other living things. It is for the latter reason that we find in them valves and veins, a uterus, umbilical vessels, and fetuses. Though it was absurd of Antoine Mizauld and Montalbani to have given plants a head, brain, heart, liver, feet, arms, skin, flesh, veins, milk, fat, nerves, glands, marrow, eyes, ears, and an esophagus, no such excesses were now in question.[60] The development of galls and cysts in plants sheds light on their formation in animals, and the success of plant grafts and the treatment of their wounds should carry over to the problems of restoring mutilated limbs and treating human injuries (p. 584). Studies of defective generation in plants should eventually shed light on the problems of infertility and abortion (p. 585).

The quarrel between medicine and microanatomy, which dragged on after Malpighi's death, repeated some familiar features of the ancient quarrel between rationalist and empiricist sects in medicine, and it mirrored a number of aspects of the quarrel occurring in England at the same time. In 1683, Gideon Harvey attacked the propensity of students of medicine to waste time "upon mangling of Piggs, Cats, Dogs, and Plucks; or upon gazing and muzling seven years upon a Hedge, Ditch, or Banks-side, to enquire for new Faces of Plants and herbs,"[61] asserting that

> the necessary point of Anatomy consists chiefly in the temperament, Figure, Situation, connexion, action and use of the parts; and not in superfluous, incertain, and probably false and indemonstrable niceties, practiced by those that flea Dogs and Cats, dry, roast, bake, parboil, steep in Vinegar, Lime-water, or *aqua fortis*, Livers, Lungs, Kidneys, Calves brains, or any other entrail, and afterwards gaze on little particles of them through a microscope, and whatever false appearances are glanced into their eyes, these to obtrude to the World in Print, to no other end, than to beget a belief in people, that they who have so profoundly dived into the bottomless pores of the parts must undeniably be skilled in curing their distempers.[62]

Harvey's attitude was not unlike that of Thomas Sydenham, a teacher of Locke, and it is reasonable to suppose that Locke's suspicion of theory, along with his suspicion of micromechanics, was influenced by the medical debates of the 1680s. Malpighi's conception of scientific activity, like Hooke's, involved a cycle of movements; from hypothesized mechanical models—in which the sieve is preeminent—to observed structures, with the hope of determining the nature of the micromachines instantiated in the body. One suspected that these little machines must be there, and endeavored to see them, finding one's expectations confirmed by, for example, the visual appearance of the glands under the microscope. The procedure was somewhat perilous, as empiricists

---

[60] For references, see Adelmann, *Marcello Malpighi*, 1:583 nn. 5 and 6.

[61] Gideon Harvey, *The Conclave of Physicians* (1686), in ibid., p. 558.

[62] Ibid., pp. 558–59.

knew.[63] The gland-sieve comparison, though technically incorrect, was at least plausible, but iatromechanism did not have other convincing examples of tiny machines visible within the organs and tissues of the body; it was more convincing in the macroscopic realm of levers and sockets. Unwilling to take on faith the deferred promise of medical micromechanics, empiricist medicine rejected as a package both instrument-assisted sense perception as a means of acquiring profound knowledge and the speculative methodology that accompanied it. It hewed to the old conception of clinical medicine, consisting of diagnostics, prognostics, and anamnestics, as a semiology.

Locke was naturally drawn into the sphere of disputes. "Seeing is believing," he said, referring to the Royal Society's replication of Leeuwenhoek's discovery of the animalcula. "We had such storys written us from Holland and laught at them."[64] But he was convinced by seeing "about 200 animals, by Dr. Wallis's computation, in the fifth or sixth part of a drop of water." In November 1678 he described the technique of making simple microscopes by melting glass beads.[65] In June 1686, he visited Leeuwenhoek in Delft and was shown some red blood cells, a human tooth, and the spermatozoa of a dog:

> I saw several of Mr. Lewinhooke's microscopicall observations which answer the descriptions he has given of them, only the globules of blood I could not clearly see, though I could see a cleare destinction between the red and pellucid parts, but the red seemed to me to lye like red threads confused laid in a clear liquor. Whether it would have succeeded any better if the bloud had been diluted with water ... wherein the red parts might have been more disjoynted I know not.
>
> The exceeding small and regular fibres of the crystalline humour are wonderfull, if all the workes of nature were not soe, as the tubes also which the bony substance of a Tooth are made up of. . . .
>
> Some of the small animalls which he said were taken out of the womb of a dog post coitum I saw sticking to a small plate of glasse. . . . They seemed to me like very small beads and twas with much difficulty I could perceive the tailes he describes of them if at least I did perceive any at all, for these being long since taken were dead and dried on to the plate of glasse.[66]

But apparently Locke had already decided that knowledge of the inner structure of bodies was of limited usefulness. This is the theme of

[63] See Duchesneau on the pitfalls of analogical arguments by iatromechanists, "Epistemological Problems of Mechanism," p. 113.

[64] Dewhurst, *John Locke*, p. 56.

[65] Ibid., p. 147.

[66] Ibid., p. 229.

the "Anatomie" fragment of 1668.[67] A minimum level of anatomical knowledge was needed, the young Locke thought, for performing some simple operations, but expertise did not provide insight into the nature of disease. "[T]hat anatomie is like to afford any great improvement to the practice of physic, or assist a man in the finding out and establishing a true method (methodus medendi), I have reason to doubt. . . . The tools where with nature works and the changes she produces in these particles [are] too small and too subtle for the observation of our senses, for when we go about to discover the curious artifice of nature, and take a view of the instruments by which she works,"[68] we are frustrated in our attempts to understand. To this he appends not a pragmatic but a principled objection: we see only surfaces. "To examine and study the parts which one dissects, without perceiving the precise mode of their functioning, is still only a superficial knowledge, and even if we dissect to the interior, we see only the outsides of things and only make a new surface to observe."[69]

Thus, even though any number of writers have seen in Locke's "new way of ideas" the origins of a distinctly modern science aiming at the correlation of phenomena—an innovation allegedly furthered by Berkeley, who went beyond Locke in rejecting unknowable corpuscular essences—it is important to recognize that there is a reactionary aspect to empiricist epistemologies. Locke's theory of the acquisition of knowledge was formulated in the climate created by Comenius and other educationalists, with their emphasis on direct experience and the collection and ordering of ideas from childhood onward. While this climate at first encouraged and supplied a justificatory rhetoric to protoscience, microscopy had outstripped it. There were many sources feeding Locke's doubt in addition to the perceived failure of contemporary microscopy. It was fed by the properly philosophical Pyrrhonian skepticism of the early seventeenth century, with the dialectic of reality and appearance converted into a dialectic of surface and interior; by Sydenham's conviction that theory is useless therapeutically; and by the disillusionment of the corpuscularians, who had failed—or so it seemed—to establish the truth of any hypothesis, and who, like Boyle, could only speculate endlessly about what might be the case. But at bottom, there was Locke's passionate belief that the natural acquisition of knowledge by children, the proper mode of educating them, and the progress that the adult human mind could make must all three be in harmony. And this entailed a move back to the surface and the natural eye.

[67] See Duchesneau, *L'empirisme de Locke*, pp. 68ff.
[68] Locke, "Anatomie," in ibid., pp. 68–69, nn. 56, 59.
[69] Ibid., p. 72.

Whereas Locke's theological views were subversive and radical for the time, on the epistemological side he was fighting a rearguard action, attempting to expand the medical conception of empiricism more generally throughout science, from human bodies to "body" in general. His project was given, for historically quite accidental reasons, an assist by Newton's phenomenological approach to the question of gravity, his refusal to speculate about the hidden causal mechanism responsible for the observable correlations between mass and attraction. It would be wrong to say that no help to science or to the formulation of scientific methods resulted from Locke's insistence that the only kind of inquiry into foundations it was possible to make was a psychological or sensualistic one. The phenomenological, statistical, correlative science that has been shown to be a genuine contribution of the eighteenth century was assisted by this flight from the microworld: such are the cunning operations of the natural dialectic.

Readers of Locke's *Essay concerning Human Understanding* are frequently tempted to ask themselves, what did Locke want? Not scholastic obscurities, not a speculative corpuscularian science on the order of Descartes's; but neither did he seem to care for the extension of the empirical horizon, the Malpighian program. To compare him with Boyle is to see essentially the same ontological scheme, in which corpuscles, textures, and primary and secondary qualities figure, but one associated with an entirely different epistemology deriving from Locke's sensationalism. Boyle, who, unlike Locke, was pious, argued that there are three worlds: a spiritual world, invisible to us, inhabited by angels and demons; a phenomenal world that is visible to us, and a dioptrical world, "which consists of all those creatures, that lay concealed, in former ages, from mortal eyes, and are not now discovered, without dioptrical glasses."[70] This is the picture that Locke undermined in his talk of analogies: he was edging away from the vision of a world in which angels and animalcula could occupy the same referential space, being merely invisible in different ways. It was still some time before the invisible in its spiritual sense was definitively cut off from the merely subvisible, and Kant could still joke in one of his late essays about whether the Leibnizian monads, spiritual substances partaking of the divine, were visible through microscopes. In Locke's *Essay*, too, the analogy of nature is said to lead us both upward, toward a consideration of higher spiritual beings, and downward.[71] But spiritual beings, though mentioned in the discourse, lie outside the discourse as the not-to-be-talked-of, which is the preliminary to their not occurring in the discourse at all.

---

[70] Boyle, *Experiments and Considerations*, in *Works* (1744 ed.), 5:773–74.
[71] Locke, *Essay concerning Human Understanding* 4.16.12.

The dioptrical world in Locke is similarly presented as the not-to-be-talked-of.

What Locke wanted, then, is fairly clear. He wanted to eliminate talk of the spiritual and dioptrical worlds and the action based on that talk in favor of religious tolerance and efficacious knowledge. Where medicine was concerned, the "historical" procedure of Sydenham struck him as exactly right: locate the phenomenal or "nominal" essence of the disease, the cluster of symptoms that characterize it and distinguish it from other diseases, and by experiment—trial and error, or, to use its more benign name, "experience"—the physician will learn the best mode of treatment, or at least the probable outcome and what to expect, wisdom consisting perhaps as much in knowledge of the latter as of the former. The physician need never make the descent to the invisible interactive mechanisms that putatively cause the illness.[72]

In his first sustained discussion of the advantages and disadvantages of enhanced optics, in the second book of the *Essay*, Locke elaborated the theme of the different ideas produced in us by a thing viewed with a microscope: "Blood, to the naked Eye, appears all red; but by a good Microscope, wherein its lesser parts appear, shews only some few Globules of Red, swimming in a pellucid Liquor."[73] This suggestion of an inferior spectacle is followed in the next section by a reminder that God has fitted our sensory organs to ordinary life and practical survival; though glimpses into underlying structure show us something of his majesty, we were not intended to have more than a glimpse. As the thin air of high mountains is unbearable to us, so exquisite hearing would be a torment. If a person's sense of sight were "1000 or 100000 times more acute than it is now by the best Microscope," Locke says, catching something of the fondness of the microscopists themselves for talk of high multiples, "things several millions of times less than the smallest Object of his sight now, would then be visible to his naked Eyes, and so he would come nearer the Discovery of the Texture and Motion of the minute Parts of corporeal things" (2.23.12). But these microscopical eyes would lead to a tragicomedy; their possessor would indeed gain knowledge, but it would be incommunicable and useless. The naturalist "would then be in a quite different World from other People: Nothing would appear the same to him, and others: The visible Ideas of every thing would be different. So that I doubt, Whether he, and the rest of Men, could discourse concerning the Objects of Sight: or have any Communication about Colours, their appearances being so

[72] See Duchesneau, *L'empirisme de Locke*, pp. 33ff.

[73] Locke, *Essay concerning Human Understanding* 2.23.11. Subsequent references to this work are cited parenthetically in the text.

wholly different" (ibid.). This point is followed by ridicule: to be able to see inside a clock to the very characteristics of the particles that give its spring their springiness would be an impressive feat, but it would not help with telling the time. To what good purpose is it if the acuteness of the owner's sense "whilst it discovered the secret contrivance of the Parts of the Machin, made him lose its use?" (ibid.). The owner would be a figure of fun at the market and the exchange (ibid.).

The impracticality of microscope eyes for ordinary life, which anyone would concede, is a different thesis from the uselessness of microscopes for scientific discovery. Yet the evident reason for conjuring up this image of optical incompetents is to throw some plausibility on the stronger thesis. For observers armed with microscopes are, however temporarily, like people with microscope eyes, residents of a world of their own, which has no common language with ours. Here Locke has a point; the discourse of the early microscope is the discourse of a traveler who reports on what others have not seen, who returns with unfamiliar descriptions of familiar objects. Our skepticism, in both cases, is tempered by our curiosity; but the combination of skepticism and love of novelty are a poor foundation for a science.

Locke got the idea for the *Essay*, as he explained in the opening "Epistle to the Reader," when, after participating in an inconclusive discussion with five or six friends in his room, he decided that there was preliminary epistemological work to be done and that "before we set our selves upon Enquiries of that Nature, it was necessary to examine our own Abilities and see, what Objects our Understandings were, or were not fitted to deal with." Once concluded, this examination revealed that the ambitions of the new science needed some restraining. Locke could not help being impressed by the arguments for the corpuscularian hypothesis reiterated from Bacon to Boyle. The hypothesis was, he conceded, "that which is thought to go farthest in an intelligible Explication of the Qualities of Bodies" (4.3.16). But he employed it only in a negative polemical context—using, for example, the results of the theory of perception allied with it as a rationale for the "new Way of Ideas."

Sometimes lack of ideational access to the realm of real essences and necessary connections was treated by Locke as an empirical limitation remediable in principle; elsewhere he wanted to draw stronger conclusions. "Microscopical eyes," he allowed at one point, might enable the observer to "penetrate farther than ordinary into the secret Composition, and radical Texture of Bodies." An observer might in this case be able to anticipate the power of rhubarb to purge, opium to stun, and hemlock to kill as easily as a watchmaker may predict malfunction in a damaged watch (4.3.25). The observer would then be able to predict in advance of experiment what new properties a known substance such as gold might have, and be on the way to achieving a true science of nature. This is not logically impossible: "Spirits of a higher rank," Locke says, "than those immersed in Flesh, may have as clear Ideas of the radical Constitution of Substances, as we have of a Triangle, and so perceive how all their Properties and Operations flow from thence" (3.11.23). He even considers the state of a spirit that might be able to adapt its organs of sight to see "the Figure and Motion of the minute Particles in the Blood, and other juices of Animals, as distinctly, as he does, at other times, the shape and motion of the Animals themselves." But he concludes that this too would probably be of no "advantage." For "God has no doubt made us so, as is best for us in our present Condition" (ibid.). Locke expressed his commitment to Boyle's middle, phenomenal world, as against the spiritual and dioptrical worlds, by saying that God had given us the appropriate faculties for our condition. He was willing to concede that a science of inner textures and operations was conceivable as a scientific horizon, only it seemed to be a drag on knowledge in the present.

> The Things that, as far as our Observation reaches, we constantly find to proceed regularly, we may conclude, do act by a Law set them; but yet by a Law, that we know not: whereby, though Causes work steadily, and Effects constantly flow from them, yet their *Connexions* and *Dependencies* being not discoverable in our *Ideas*, we can have but an experimental Knowledge of them. From all which 'tis easy to perceive, what a darkness we are involved in, how little 'tis of Being, and the things that are, that we are capable to know. . . . as to a perfect *Science* of natural Bodies, (not to mention spiritual Beings,) we are, I think, so far from being capable of any such thing, that I conclude it lost labour to seek after it. (4.3.29)

Some commentators have argued that book 4 of the *Essay* shows that Locke was not opposed to the employment of hypotheses concerning unobservable events and processes and that he wished only to stress that corpuscularian explanations and the micromechanical or microchemical explanations of theoretical medicine were tentative and could never be accepted as certain or necessary.[74] Certainly he does not seem to have doubted that light and heat were equivalent to patterns of underlying physical activity, though it was in keeping with his general position to refuse to align himself with any particular theory. However, there is a difference between an open-minded agnosticism about the choice of a micromodel to explain a given effect and Locke's view that

---

[74] See, on this debate, Laudan, "Locke's Views on Hypotheses"; Yost, "Locke's Rejection"; and Yolton, *Locke*, pp. 59ff.

seeking a microscopical science of bodies is "lost labor" and that progress in all practical fields can be accomplished better without it. Anyone who doubts that Locke's view was, in the context of late-seventeenth-century science, deeply skeptical and pessimistic might consult the statements in his treatise *Some Thoughts concerning Education* on the teaching of natural philosophy. Here he repeats that natural philosophy will never become a science. "The Works of nature are contrived by a Wisdom, and operate by ways too far surpassing our Faculties to discover, or Capacities to conceive, for us ever to be able to reduce them into a Science."[75] "I do not," he says, "conclude that none of them are to be read: It is necessary for a Gentleman in this learned Age to look into some of them, to fit himself for Conversation." But the aim of doing so is only to become conversant with various hypotheses, rather than to try "to gain thereby a comprehensive, scientifical, and satisfactory Knowledge of the Works of Nature."[76] There is much that is "convenient and necessary to be known to a gentleman," but those who have made experiments and observations and those who merely recount them were to be preferred over systematists. "Such writings therefore, as many of Mr. Boyle's are, with others that have writ of Husbandry, Planting, Gardening, and the like, may be fit for a Gentleman."[77]

On the whole Locke impresses us as one somewhat bemused by the science of his time. He realized that great things were happening, things that tended to reduce the importance of philosophy as traditionally conceived, and he anticipated this dethronement by offering his services as a factotum. Not everyone could be a "Master-Builder" on the order of Newton, Huygens, Sydenham, or Boyle, but "'tis Ambition enough" in his view "to be employed as an Under-Labourer in clearing Ground a little, and removing some of the Rubbish that lies in the way to Knowledge."[78] But was he helping, or hindering, or both? Part of the reason for his ambivalence toward the new science was that Locke was still faithful, in certain ways, to an older image of knowledge. The ideal against which he measured the new micromechanical science of bodies and found it wanting was mathematical demonstration, in which agreement of ideas is secured through logical connection—through entailment relations, as we would say. Locke wielded his epistemological criticism against metaphysicians such as Descartes, with his "thinking substance," who did not seek a secure grounding for their concepts in

[75] Locke, "Some Thoughts concerning Education," in *Educational Writings*, p. 301.

[76] Ibid., p. 305.

[77] Newton, however, is an approved author because he does prove his conclusions about the solar system by advancing mathematical arguments; ibid., pp. 306–7.

[78] Locke, *Essay concerning Human Understanding*, epistle to the reader, pp. 9–10.

experience; yet when it came to formulating an epistemology of experimental science, the same distrust of the conjectural caused him once again to retreat to the level of sensory immediacy.

And Locke was troubled by disproportion. If our interests had really depended on a knowledge of the microworld, a benevolent God would not have put it out of our reach; this thought was shared by the other great empiricist and occasional microscope-skeptic, George Berkeley. Berkeley thought more deeply and ultimately better about the science of his time than Locke did; yet the same existential concerns were there. How can it be that what is vital to our health and well-being is all taking place at a level where we cannot see or cannot see well? Must there not be a harmony between what our God-given faculties show us and what it is good for us to know?

Berkeley referred in a number of places to microscopes. His most famous statement was an echo of Hooke's claim that the microscope shows us a new world, that "every considerable improvement of *Telescopes* or *Microscopes* [produces] new Worlds and *Terra Incognita's* to our view."[79] Only he took this to mean that we do not see the same world, made up of the same objects, more clearly with a lens, but entirely other objects. And this is indeed a conclusion one might derive from leafing through the *Micrographia*. We are told there, because we could not otherwise guess it, that what we are seeing are bits and fragments of familiar substances—moss, mold, and vermin—presented under an unfamiliar aspect. Mold looks like a lunar forest, a flea looks like a crustacean, and so on. We might as well be seeing new landscapes and new forms of life; only their labels identify them. And from this one might then infer that the microscope can give us no information—or at least no direct information—about our macroscopic world and its objects, except information of the general form "this is how it looks under the microscope."[80] Yet Berkeley's attitude toward the microscope and the existence of the microworld was not one of simple repudiation.[81] He did not hesitate to avail himself of arguments for perceptual relativity based on the changes in close versus long views, as though the microscope had a positive role to play in philosophy, nor did he deny that there are events that take place beneath the threshold of ordinary vision. He agreed that animal and plant bodies are micromachines. In *The*

---

[79] Hooke, *Micrographia*, preface. Elsewhere in the preface he says that "by the help of *Microscopes*, there is nothing so *small*, as to escape inquiry; hence there is a new visible World discovered to the understanding." Grew's dedication in the *Anatomy of Plants* also refers to new worlds.

[80] Cf. Berkeley, *An Essay towards a New Theory of Vision*, in *Works* 1:205; *Three Dialogues*, in *Works* 2:245.

[81] See Brykman, "Microscopes and Method in Berkeley."

Principles of Human Knowledge (1710), he praised "the curious organization of plants, and the admirable mechanism in the parts of animals ... and all the clockwork of Nature, the greater part whereof is so wonderfully fine and subtle as scarce to be discerned by the best microscope."[82]

Commentators have been rightly puzzled by this praise, thinking that it jibes neither with Berkeley's opposition to corpuscularian theories nor with his doctrine that *esse est percipi* and that we perceive only our own ideas and not material things. If we take his acceptance of clockwork animals at face value, then Berkeley does not look very different from his supposed opponents, and one begins to wonder whether his idealism was anything more than a sort of recommendation for linguistic revision. If a philosopher accepts the existence of invisible microentities that have a productive role in bringing out effects at the surface level, we may wonder what the force of his saying that there are no active powers in nature, no causal production, and no substances apart from particular minds is. Is this not simply a decision to call objects "collections of ideas," and to call causal sequences "regular correlations"? In that case Berkeleian immaterialism is merely a sort of operator that can be prefaced to any ontology and any account of the world at all, with a slight effect on the sense of what is said but no effect on the truth value. But the way to make sense of Berkeley's views about the microworld and his youthful philosophy generally—and this means supposing that his metaphysics was substantive and not lexical—is to keep firmly in mind what it was that troubled him about the microworld and its relation to ordinary experience, what theses he would necessarily reject about causation and explanation.

Note first in this connection that a philosopher can accept some form of mechanism without accepting corpuscularianism. In the *Notebooks*, Berkeley inveighed against corpuscularians, among whom he included Epicureans, Spinozists, Cartesians, Locke, and Boyle. "The sillyness of the Currant [*sic*] Doctrine makes much for me. They commonly suppose a material world, figures, bulks of various sizes etc according to their own confession to no purpose, all our sensations may be and sometimes actually are without them. Nor can we so much as conceive it possible that they should concur in any wise to the production of them."[83] The common use of the "clock metaphor" by corpuscularians to elucidate the relation between inner motions and configurations on the one hand, and the existence of qualities and the production of effects at the surface level on the other, smoothed over two problems that

[82] Berkeley, *Principles*, in *Works* 2:60–66.
[83] Berkeley, *Notebook*, A 476, in *Works* 1:60.

Berkeley registered here and elsewhere. A naive observer who looked at a clock face would not deduce that a particular system of toothed wheels and springs was responsible for the hands going around the dial, and a naive observer who saw only the toothed wheels would not guess that this was the inner mechanism of a clock. Thus the relation of the face to the inner structure is discovered only a posteriori, and the face can exist without the mechanism and vice versa. But if you have this structure, then, ipso facto, you have a clock: the experienced observer who is permitted to examine the dial, the back of the dial, and the clockwork understands why a mechanism like this tells the time and why only some roughly similar mechanism will do the same thing. How the connection between inner configuration and surface could be both discoverable only a posteriori and necessary was clearly puzzling.

If we allow for the necessary a posteriori, another problem looms up: what happens when we cannot understand the connection between the face and the internal structure? Leibniz pointed out the problem in the famous "mill" passage of the *Monadology*: no knowledge of internal operations explains the consciousness of qualities.[84] A knowledge of the microstructure and microactivity of acids can be supposed to explain why acids etch metals or dissolve solids. But a knowledge of the microstructure and microactivity of the corpuscles that supposedly cause us to see green cannot help us understand how green is seen. This is the problem of the "arbitrariness" of the relation between secondary properties and their corpuscularian causes that Locke worried about in book 4 of the *Essay*, finding "no conceivable connection" between the size, figure, and motion of particles and the colors, tastes, and sounds supposed to depend on them. He decided that we could reason about them only as "effects produced by the appointment of an infinitely Wise Agent which perfectly surpass our comprehension."[85]

In Boyle's early paper on colors, he had cited reports of people who were able to discriminate different hues by feeling them with their fingertips as evidence for his theory that color is a feature of the surface texture of an object. His hope was that the microscope would enable us to discern the "latent ruggednesses," the "little protuberances and cavities," that "interrupt and dilate" the light and determine the color we see.[86] "[W]ith an excellent microscope," he said, "where the naked eye did see but a green powder, the assisted eye . . . could discern particular granules some of them of a blue, and some of them of a yellow color."[87] So one might have understood him as claiming that the microscope

[84] Leibniz, *Monadology*, 17, in *Philosophische Schriften* 6:609.
[85] Locke, *Essay concerning Human Understanding* 4.3.28–29.
[86] Boyle, *Experiments and Considerations*, in *Works* (1744 ed.), 5:680ff.
[87] Ibid., p. 680.

would explain how we see colors. But although the microscope may show us how colors combine to form new colors, it cannot show us, in a radical sense, the origin of colors. And what after all has the tactile impression of roughness to do with color? The microscope and micro-analysis, in Berkeley's view, simply encourage us to do bad philosophy. Newton himself was forced into obscure language when he said that sound is really a trembling motion of the air that only takes the "form" of a sound in the sensorium, and that colors are a disposition of an object to reflect a certain collection of rays that then cause the "sensations of those motions under the form of Colors."[88]

Thus Berkeley's position included and went beyond Locke's in rejecting arbitrary, unexperienced connections. It is not only lost labor to seek a corpuscular science of bodies that will never explain perception; hypothesized corpuscular essences, having had no useful, predictive role to play so far in science, do not deserve retention. Locke's statement that secondary properties are the appointments of an infinitely Wise Agent that we cannot comprehend is, on Berkeley's reading, another sly statement by a religious skeptic who wants to associate God with unintelligibility. Nevertheless, there was no reason why, for Berkeley, an anticorpuscularian stance should preclude an interest in anatomy and in the bodily machine. We can make sense of the notion that the gross anatomy of the animal body—especially the heart and circulatory system—elucidates the vital phenomena of warmth and motion. Science can also develop a system of diagnostic and predictive signs based on the appearance, even perhaps the microappearance, of glands, tissues, and organs, and with autopsies one may retrodict the cause of death. But all this has nothing to do with unseen corpuscles and corpuscularianism.

This was Berkeley's actual position in his *Three Dialogues* and in *The Principles of Human Knowledge*. The plant really is a machine, but the microscope will never show it to us as such: a microsection of a plant stem is, as Grew feared, scientifically quiet, a complex and nice design. The microscope gives us only pictures or views; it cannot show us how a plant blooms, poisons, or reeks. And if the micrographical picture is a picture on the way to the picture that the corpuscularian would like to be able to draw, we have no reason for confidence that some picture even more remote from our ordinary experience will suddenly prove to be richly explanatory.

Mechanical philosophers, on Berkeley's view, imagine that the inner texture and motion they see can actually bring about the life of the plant, but it is superstitious to believe of an image that it is efficacious.

---

[88] Newton, *Opticks*, book 1, part 2, prop. 2, theorem 2, definition, p. 125.

Representation freezes the image of the living or formerly living thing. It is as though one were to point at one of Grew's cross-sections and command it to grow. But even if Berkeley had thought us able to see the flow of sap in the vessels, he would have protested that we were not seeing its life. God, as he pointed out, might have created hollow plants and animals that lived, moved, grew, and flowered without benefit of any interior structure. The science of Berkeley's early work, then, can be described as one informed by a micrographical awareness, but it is one in which nothing essential or important occurs at the edge of our perceptual world, or more exactly, in those worlds that the microscope shows us. Microscope eyes could afford us at best, he tells us in the *New Theory of Vision*, the "empty amusement of seeing."

The theory of divine visual language that Berkeley offered instead of a science of inner textures and motions was an assertion of the meaningfulness of immediate visual experience: the world shows us the "production and reproduction of so many signs, combined, dissolved, transposed, diversified, and adapted to such an endless variety of purposes."[89] The correlationist science that Locke had expropriated from medicine to chemistry was expanded, so that the statement of regularities in experience rather than the discovery of internal productive mechanisms was the aim of research. Nature teaches us how to act and what to expect. Yet later in life Berkeley too submitted to the new science. The work of his old age, *Siris*, is a veritable treatise *de subtilitate*, abounding in references to minute fibers, filaments, corpuscles, particles, and vibrations existing far beneath the threshold of perception and responsible for our feelings of vitality and debilitation: the microworld is not only thought, it is thought as active and productive.[90]

The late eighteenth century saw a strong reaction against the system of organic molecules and the theory of preformation, which were regarded as examples of microscope-induced delusion.[91] Both the error of sponta-

---

[89] Berkeley, *Alciphron*, in *Works* 3:159–60.

[90] See Wilson, "Berkeley and the Microworld."

[91] "The author of the *System of Nature*," Jean Sénebier (1742–1809), writes, referring to Buffon, "has strangely abused the revelations of the microscope" in arguing from the presence of animalcula in infusions to a *vis essentialis* present in matter. *Opuscules de physique animale et végétale, par M l'abbé Spallanzani* 1:lxvi, in Marx, "L'art d'observer au XVIII siècle," p. 209. Antoine Serres (1786–1868) casts the blame even further back, observing that it was the abuse of the microscopical researches of Leeuwenhoek and Hartsoeker that led physiologists into the snare of preformation theory. "First one saw the animal inside the egg put to incubation, then one saw it in the egg before conception, and finally . . . Swammerdam and Malebranche imagined the preexistence of germs and their emboitement from eternity. These monstrous ideas had a success which the discoveries of Galileo and Newton never equalled." *Principes d'embryogénie, de zoogénie et de teratogénie* (1860), in Marx, "L'art d'observer au XVIII siècle," p. 209.

neous generation, attacked by preformationists, and the error of preformation, attacked by epigenesists, as well as the excesses of worm theory, were laid to the deceptive nature of the instrument. But the old dogma that technical improvements in lenses, especially the solution of the problem of chromatic aberration, were responsible for its valorization cannot be accepted. A. F. W. Hughes, who investigated this claim in 1955, determined that "the eighteenth-century microscopist could have had a better instrument had he really wanted one and ... the factors which limited progress at this time were not really technical in nature."[92] Progress in botany was slow because no observers were even describing clearly what could be seen with the naked eye or with low-power microscopes, and the pace of research and its success (for example, in embryology) quickened in the early eighteenth century even before the introduction of achromatic lenses, which were rather a response to new demand than a condition of productive investigation. Leeuwenhoek's representations of spermatozoa were as accurate as any produced with the early achromatic microscope. Fournier's more-recent study reinforces the claim that technical improvements and inadequacies did not account either for the rise of experimental microscopy or for its decline after 1680 and again after 1750.[93] Although embryology was on a better footing in the first quarter of the nineteenth century, both plant studies and pathology appeared to be infected by despair. Augustin Pyramus de Candolle, a botanist, wrote in 1827 of the "incertitude désespérante pour les amis de la verité"; Xavier Bichat (1771–1802), a genius of observation who classified tissue with the naked eye, returned to a kind of Harveian primacy of direct experience when he banned the microscope from the pathology laboratory as giving entirely subjective results: "chacun voit a sa manière et suivant qu'il est affecté."[94] In the first years of the nineteenth century, according to Emile Blanchard, savants were so unimpressed by the microscope that they even denied the existence of blood corpuscles. Definitive acceptance occurred much later, but Rudolf Virchow reported in 1855 that "the promotion of microscopical research, initially mocked, is now victorious, and the language and concepts of pathological anatomy, indeed anatomy in general, are based in cellular pathology."[95]

The eventual establishment of the microscope as a trustworthy and informative research tool was, then, not a matter simply of its technology but rather one of its context. Ronchi's claim that optical instruments were accepted as providing a reliable route to new knowledge only when the theory of their functioning was properly understood ap-

[92] Hughes, "Studies in the History of Microscopy."

[93] Fournier, "Fabric of Life," pp. 187ff.

[94] See Marx, "L'art d'observer au XVIII siècle."

[95] Berg and Freund, *Geschichte der Mikroscopie*, p. 11.

plies better to their rehabilitation than to their initial habilitation in the seventeenth century. Renewed insistence and demonstration that both the eye and the microscope functioned according to immutable mathematical laws was an element in this rehabilitation.[96] But so too was the insistence that more than a good eye and a good microscope were required for the kind of seeing that produced knowledge: skill, familiarity, even a kind of moral virtue entered into the picture of the competent observer that emerged fully in the nineteenth century.[97] It will be profitable to try to understand this development by determining how microscopical images achieved stability—how descriptions were confirmed, how images came to be accepted as accurate, how potentially distorting techniques of preparation came to be accepted as revealing rather than as destructive of the specimen. Such a diachronic study will require close attention to the language of microscopy, its institutional supports, the character of its practitioners and of their interaction with one another, and microscopy's reaction to internal and external criticisms.

[96] See, on M. J. Schleidan's role in this respect, Chadarevian, "Art of Experimenting."
[97] On observership and morality, see Daston and Galison, "The Image of Objectivity."

8

Truths and Appearances

THE PREVIOUS two chapters have documented the sense of dislocation induced by the discovery of the microworld, an effect balanced by the strongly positive reception of the microscope by those enthusiastic philosophers who were enchanted by images of an infinite complexity, regularity, and variety of animate and inanimate forms in nature. The consistent presence of the microworld in the philosophical literature of the period and the welcoming or rejecting attitude attached to it is proof that, for late-seventeenth-century natural philosophers, the sudden movement outward of the perceptual horizon could not be a matter of indifference. As George Bernard Shaw said, when you have learned something it feels as though you have lost something, and their repeated expressions of incapability and doubt are the best possible evidence that they were learning something. Support of one's ideologies when it appears from unexpected quarters is, by contrast, always a matter for rejoicing, and in holding at bay some of the more ominous forms of materialism and mortalism, the microscope served some human interests as it undermined others.

It is appropriate here to consider Gaston Bachelard's assessment of the microscope. It was his view that it constituted an actual impediment to knowledge in the seventeenth and eighteenth centuries, because it revealed things that were beautiful but that could not lead to the formation of new theories and things that suggested to observers theories of the wrong sort—about the organic complexity and vivacity of inanimate nature—and that made a quantitative handling of the phenomena seem impossible or unimportant. "Its decisive valorization," he argues, "lay in the discovery of the hidden under the manifest, the rich under the poor, the extraordinary under the ordinary. . . . In truth, it was only a case of spinning out the old dreams with the new images which the microscope delivered. That people sustained such excitement over these images for so long and in such literary form is the best proof that they dreamed with them."[1] In Bachelard's construction, the introduction of the microscope was a matter of bad timing. Its first appearance was within a science still permeated by subjectivity, by, despite its rhetoric, dream and hope, and these elements were expunged only gradually

[1] Bachelard, *La formation de l'esprit scientifique*, p. 160.

from a science that still drew off the emotional valences of its subject matter.

One argument of the present study has been that the microscope, like other instruments and devices, brought into being new phenomena. These phenomena were dreamed about, to be sure, and they were in some cases as fleeting and insubstantial as dreams, but they had the intractability of fact, and when they could not be fit into preexisting systems, then systems were made around them. If, as Marian Fournier argues, physiological inquiry shifted in the early eighteenth century "from experimental and ocular investigation towards mathematical and rational analysis,"[2] the problem was not that visual illusion and psychological projection had led to dead ends. Microscopy stalled because—the initial mapping of the major plant and animal organs and tissues as seen through the microscope completed—there seemed, as Hooke said, nothing more to be done. This mapping had been accomplished, but it had neither brought an understanding of process nor increased the potential for beneficial intervention into the workings of the animal machine. The macroscopic, quantitative approach that was first applied to plant physiology and the study of animal heat was a retreat to the surface and to problems that were less recalcitrant than that of explaining function by structure. The concreteness and emphasis on physical engagement with living things that this new approach borrowed from the observational and experimental sciences set it apart from philosophical or speculative mechanism, which first existed in a state of reciprocal encouragement with microscopy but was ultimately abandoned by the life sciences. The "mechanization of the world picture," in the celebrated phrase of Eduard Dijksterhuis, was thus interrupted by microscopy and reconstituted in new terms.

Bachelard is right to emphasize that human interest was in a constant state of interaction with the new data of microscopical observation, that the unfamiliar and incomprehensible were tamed and made intelligible by being described and theorized in the language and concepts of the world of immediate lived experience and affect. But the development of modern science did require that observation and description should be divested of immediate reference to the pleasurable, the meaningful, and the useful, and the "new" and largely useless world of the microscope lent itself to that task as much as it invited comparison and elicited both mundane and ecstatic emotion. The distance from the subject and appreciation of form in the abstract of the authors of the earliest works on microscopy was not, in and of itself, new: the ancient writers on medi-

---

[2] Fournier, "Fabric of Life," pp. 196–97.

cine, botany, agriculture, and embryology had it, and the mood informing Marcus Aurelius's contemplation of the cracks in baked bread for their look alone, and not for the moral lesson contained in them, is not so different from Kepler's meditation on the formative principles of living and inanimate nature in the snowflake. But consciousness of form reached new extremes in the literature with which I have been concerned. Malpighi was struck by some patterns of frost that he discovered formed only on the windowpanes of newly plastered rooms. He observed them under the microscope and sketched their "angular frustules," observing how they resembled plants.[3] We know now that he—and before him, Harvey, who saw the umbilical vessels of the three-day-old chicken embryo as resembling tree branches—was responding to the beauty of fractal structures, the reiterated shapes that are the building patterns of nature and that baffled Kant, because he saw that nature seems to create them effortlessly, that we cannot create them, and yet that they seem to be meant for our eyes.[4] Malpighi saw them in that supremely ignoble object, the placenta:

> The other part of the gland, which is adnate to the chorion is made up of frayed roots, so to speak, which enter the little sheaths I have described, and you will see a lovely sight when you examine this part with the microscope after it has been withdrawn from the sheaths and kept in water for a long time, when the separated rootlets and capillaments gradually lift up so that they resemble an exquisite forest; for innumerable roots like those of plants arise from a common trunk, divide into a multiplicity of small branches, and fray out into hairlike filaments.[5]

To return to Bachelard's thesis, there were indeed a number of ways in which people dreamed with the new images presented by the microscope. They dreamed in that they could experience them often in only a passive way, unable to put what they saw to further use. They dreamed in assigning them a place in fanciful, premature, or conceptually unsound explanatory programs. And they dreamed in spinning around the instrument a web of moral and theological significance, supposing that they saw with it traces of the delineations of the invisible finger of God. Hooke had forgotten or suppressed much of his knowledge of what the microscope had been used for in his *Discourse* of 1692; when later researchers wanted to defend the scientific importance of the instrument by contrast with its recreational interest, they tended to point over and over to the same example: the direct verification of the circulation of the blood in animals with diaphanous blood vessels such as those in fish

[3] Adelmann, *Marcello Malpighi* 1:420.
[4] Kant, *Kritik der Urteilskraft*, in *Gesammelte Schriften* 5:348.
[5] Ibid. 1:434.

and frogs. It would be a long time before anyone could see with the instrument anything that might help to make that viewer, in the sense of Bacon and Descartes, a master and possessor of nature.

As technically limited and as amateur-infested as early microscopy was, and as skeptically as its promise was sometimes viewed, the knowledge it produced was nevertheless real. Among the positive theoretical achievements of seventeenth- and early-eighteenth-century microscopy one can list, following Schrader,[6] knowledge of the insect body and its organs, the collection of strong evidence against the spontaneous generation of insects, the discovery of a capillary system in plants, the discovery of the gradual stages involved in insect metamorphosis and the growth of the fetus, the discovery of the characteristic shapes of crystals, and the discovery of infusoria. In many areas the microscope's testimony was ambiguous. It showed that the harmonies of form were ubiquitous, but it also indicated that there was a line separating the animal and vegetable worlds, in which generation and growth proceeded through similar processes, from the mineral world. Numerous details were added to the animal machine. Besides the capillary structure that completed the circulation of the blood, the textures of glands, tissues, and organs were repeatedly described.

Even in its production of a new micrographical literature, the microscope gave sense to the claim that the new science was nondiscursive, that the subtlety of nature was not the same as the subtlety of the mind. Yet a mere mapping, a mere delineation of what was seen, was not knowledge; a representation was not a theory. And so I return to the question posed in chapter 7: in what sense did the images provided by the microscope explain anything?

First, in a broad sense, the microscope put into question that type of explanation that required appeal to formal and final causes, especially the replicative causation in which an idea or a formal immaterial agent produces a copy of itself. Certainly the epigenesis and the vitalism of the eighteenth century might seem to be restorations of final and formal, as opposed to efficient and material, causality. But these were quite different from the animistic theories of contagion and generation of Harvey and Helmont. Second, the microscope substituted subvisible textures and motions for occult powers, dispositions, and capacities, and so gave meaning to the notion of the interpretation of nature that neither the corpuscular philosophy of Descartes and Boyle nor the mathematical philosophy of Galileo and Newton had been able to supply. The all-around excellence of the mechanical hypothesis notwithstanding, it was as remote and hypothetical in the programmatic writings of Descartes

---

[6] Schrader, *De microscopiorum usu.*

and Boyle as it had been in those of Lucretius. The grip that it held in the early modern period was related to the promise of actually being able to see, with the help of the microscope, tiny machines and invisible processes. Microscopical observation promised to give a deeper and more profound view of the world, but one that was based in experience rather than thinking. It offered to reconstitute the discipline and communality of sense experience without its problematic superficiality. And it would provide a rival praxis for natural philosophers to that of textuality and conceptualization, citation and demonstration.

But why should we admit that the image delivered by the microscope is a better image, or a deeper or truer image, rather than simply another image? Here we encounter the paradox formulated by Mersenne, by Locke, by Fontenelle, and by Hume: We see only surfaces, whatever we see is the effect of hidden machinery; therefore whatever we see is scenery and not truth, and, as an illusionistic object, it can play no causal role. Microscopy generates representations, but representations by themselves do not explain anything. Does this not show that a deep observational science of nature is impossible, that the problem is not, as Fontenelle frivolously suggested, that our curiosity outruns our eyesight, but that there is an antinomy of surface and interior that the pragmatic success of microscopy has not solved, and that our rejection of the extreme solution proposed by Berkeley does nothing to mitigate?

It is not a full solution, though it is certainly true, to say that the microscope can show us processes, not merely successive states. The worry of the philosopher is rather that what we see is always a product—an idea, a visual image—rather than something productive, so that it seems to follow that, if there are truly productive entities in nature, we cannot see them. But to admit that what we see in inspecting the details of a cell membrane is an appearance produced by some deeper reality that is not simultaneously visible is not to deny that what we observe there might explain a process of coarser grain, such as the healing of a wound. That the construction of a stage set explains the appearance of a drawing room is not to be doubted, even if that stage set itself is only an appearance produced by an underlying arrangement of molecules, and so on. The conclusion to which this seems to lead is that even reductive "explanation" is itself a relation holding between appearances. And this seems at first a curious discovery, for appearances do not seem to have the weight and push, or the invariability, of reality.

There can be, however, no regress to the infinite on this account. The existence of an optical limit of resolution entails that certain objects of a mathematically determinable degree of smallness cannot be seen with a light microscope. Other physical constraints of nature will determine that other entities of an assignable degree of smallness are not visible by means of other imaging techniques. It is thus strictly correct to say that our theories do not represent, do not correspond to, reality, and yet it is true that there is a reality beyond our representations and that it is the weight and push of that reality that determines the appearances for us. It is true, then, both that scientific apparatuses permit us to see more of the world as it really is and that they produce for us only an illusory image, which we can learn nevertheless to turn to practical ends. The image of natural optics is the result of filtering and processing, comparing and averaging, and is simply that function of the available light energy that has proven to be minimally adequate for survival over the course of evolution. It cannot be directly compared with what is "there." Our familiar commerce with the visual world, which now gives us the paradigm of clear, distinct, immediate perception, had in our infancy to be negotiated out of ambiguity and confusion. We can alter the resulting function by placing an artificial lens in front of the natural one. And although we seem at first thereby to be extending the domain of what can be clearly, distinctly, and immediately perceived, we are only forced again to extract meaning from the optically indeterminate.

Bibliography

"An account of a book intitled *Ricreatione dell' occhio e della Mente Nell' osservation delle Chiocciole* Dal P. Filippo Buonanni &c. in Roma, per il Varese, 1681." *Philosophical Transactions* 14, no. 156 (1684): 507–9.

"An Account of the Dissection of a Bitch . . . by an ingenious Physitian." *Philosophical Transactions* 13, no. 147 (1683): 187.

Adelmann, Howard. *Marcello Malpighi and the Evolution of Embryology*. 5 vols. Ithaca: Cornell University Press, 1966.

Agrippa von Nettesheim, Heinrich Cornelius. *De occulta philosophia*. 1533. Reprint, Graz: Akademische Druck u. Verlagsanstalt, 1967.

———. *The Vanitie and Uncertaintie of the Artes and Sciences*. 1531. Translated by J. Sanford. London, 1575.

Alpers, Svetlana. *The Art of Describing*. Chicago: University of Chicago Press, 1983.

Andry de Boisregard, Nicolas. *An Account of the Breeding of Worms in Human Bodies*. London: Rhodes and Bell, 1701. Originally published as *De la génération des vers dans le corps de l'homme* (Paris, 1700).

"Another Extract Out of the Italian Journal [*Giornale de letterati*], being a Description of a Microscope of a New fashion, by the means whereof there hath been seen an Animal lesser than any of those seen hitherto." *Philosophical Transactions* 3, no. 42 (1668): 842.

Aquinas, Saint Thomas. *Summa theologica*. 60 vols. New York: McGraw-Hill, 1964–67.

Aristotle. *The Generation of Animals*. Translated by A. L. Peck. London: Heinemann, 1943.

———. *The History of Animals*. Translated by A. L. Peck. London: Heinemann, 1965.

———. *The Parts of Animals*. Translated by D. M. Balme. Oxford: Clarendon Press, 1972.

Augustine, Saint. *The Trinity*. Translated by Stephen MacKenna. Washington, D.C.: Catholic University of America Press, 1963.

Bachelard, Gaston. *La formation de l'esprit sciéntifique*. 8th ed. Paris: J. Vrin, 1972.

———. *La poétique de l'espace*. Paris: Presses Universitaires de France, 1957.

Bacon, Francis. *Works*. Edited by James Spedding, Robert Leslie Ellis, and Douglas Denon Heath. 15 vols. Cambridge: Riverside Press, 1857–61.

Bacon, Roger. *The Opus majus of Roger Bacon*. Translated by Robert Belle Burke. 2 vols. Philadelphia: University of Pennsylvania Press, 1928.

Baker, Henry. "An Account of Mr. Leeuwenhoek's Microscopes; by Mr. Henry Baker, F.R.S." *Philosophical Transactions* 41, no. 458 (1739–40): 503–19.

———. *An Attempt towards a Natural History of the Polype*. London: R. Dodsley, 1743.

——. *The Microscope Made Easy.* 2d ed. 1743. Reprint, Lincolnwood, Ill.: Science Heritage, 1987.

Balss, Heinrich. "Praeformation und Epigenese in der Griechischen Philosophie." *Archeion* 4 (1923): 319–25.

Barer, R. *Lecture Notes on the Use of the Microscope.* Oxford: Blackwell, 1968.

Bassi, Agostino Maria. *Del mal del segno, calcinaccio o moscardino.* 2 parts. Lodi, 1835–36.

Bayle, Pierre. *Dictionnaire historique et critique.* 2d ed. 3 vols. Rotterdam, 1702.

——. *Oeuvres diverses.* 1737. Reprint, 4 vols. in 5, Hildesheim: Olms, 1964–82.

Bell, Louis. *The Telescope.* New York: McGraw-Hill, 1922.

Belloni, Luigi. "Appunti per una storia pre-Leeuwenhoekiana degli 'animalcula.'" *Gesnerus* 23 (1966): 13–22.

——. "Athanasius Kircher: Seine Mikroskopie, die Animalcula und die Pestwürmer." *Medizinhistorisches Journal* 20 (1985): 58–65.

——. "Charlatans et *Contagium Vivum* au declin de la première période de splendeur du microscope." In *Comptes rendues du 85ième Congrès des sociétés savantes,* pp. 579–87. Paris, 1961.

——. "Dalla microscopia alla anatomia microscopica dell'insetto." *Clio medica* 4 (1969): 179–90.

——. "De la theorie atomistico-mechaniste à l'anatomie subtile (de Borelli à Malpighi) et de l'anatomie subtile à l'anatomie pathologique (de Malpighi à Morgani)." *Clio medica* 6 (1971): 99–107.

——. "Malpighi and the Founding of Anatomical Microscopy." In *Reason, Experiment, and Mysticism in the Scientific Revolution,* edited by Maria Luisa Righini-Bonelli and William R. Shea, pp. 95–110. New York: Science History Publications, 1975.

——. "Micrografia illusoria e 'animalcula.'" *Physis* 4 (1962): 65–73.

——. "Schemi e modelli della macchina vivente nel seicento." *Physis* 5 (1963): 259–98.

Berg, Alexander, and H. Freund. *Geschichte der Mikroskopie.* Frankfurt am Main, 1963.

Berkeley, George. *The Works of George Berkeley.* Edited by Arthur Aston Luce and Thomas Edmund Jessop. 9 vols. London: Thomas Nelson and Sons, 1948–57.

Berthier, A. G. "Le mécanisme cartésien et la physiologie au XVII siècle." *Isis* 2 (1914): 37–89.

Birch, Thomas. *The History of the Royal Society.* 4 vols. 1756–57. Reprint, Hildesheim: Olms, 1968.

Blanchard, Emile. "Les premiers observateurs au microscope." *Revue de deux mondes* 76 (1868): 379–416.

Blumenberg, Hans. *Die Lesbarkeit der Welt.* 2d ed. Frankfurt am Main: Suhrkamp, 1983.

Boehme, Jakob. *De signatura rerum.* Frankfurt am Main: Insel, 1980.

Boettger, C. F. *Foetum non ante conceptionem in ovulo praeexistere, sed post eandem formari.* Leipzig, 1708.

Boghurst, William. *Loimographia: An Account of the Great Plague of London in the Year 1665.* Edited by J. F. Payne. London: Epidemiological Society of London, 1894.

Bonnet, Charles. *Considérations sur les corps organisés.* Amsterdam, 1762.

———. *La palingénésie philosophique.* 2 vols. Geneva: C. Philibert and B. Chirol, 1769.

Bonomo, Giovanni Cosimo. *Osservazioni intorno a' pellicelli del corpo umano.* Florence: Martini, 1687.

———. "An Abstract of . . . a Letter . . . containing some Observations Concerning the Worms of Humane Bodies." *Philosophical Transactions* 23, no. 283 (1702–3): 1296–99.

Bordo, Susan. *The Flight to Objectivity: Essays on Cartesianism and Culture.* Albany: State University of New York Press, 1987.

Borel, Pierre. *De vero telescopii inventores . . . accessit etiam centuria observationum microscopicaricum.* The Hague, 1655–56.

Boyle, Robert. *Works.* Edited by Thomas Birch. 5 vols. London: A. Millar, 1744.

———. *Works.* Edited by Thomas Birch. 6 vols. 1772. Reprint, Hildesheim: Olms, 1965–66.

Bracegirdle, Brian. *A History of Microtechnique.* Ithaca: Cornell University Press, 1978.

Bradbury, Savile. *The Evolution of the Microscope.* Oxford: Pergamon, 1967.

———. *The Microscope, Past and Present.* Oxford: Pergamon, 1968.

———. "The Quality of the Image Produced by the Compound Microscope: 1700–1840." In *Historical Aspects of Microscopy,* edited by Savile Bradbury and Gerald L'E. Turner, pp. 151–72. Cambridge: Heffer and Sons, 1967.

Bradbury, Savile, and Gerald L'E. Turner. *Historical Aspects of Microscopy.* Cambridge: Heffer and Sons, 1967.

Bradley, Richard. *New Improvements of Planting and Gardening.* London: W. Mears, 1718.

———. *The Plague at Marseilles Considered.* London: W. Mears, 1721.

Bronn, Heinrich Georg. *Klassen und Ordnungen des Thier-Reichs.* 5 vols. Leipzig: Carl Winter, 1866–1919.

Browne, Thomas. *Selected Writings.* Edited by Geoffrey Keynes. Chicago: University of Chicago Press, 1968.

———. *Works.* Edited by Geoffrey Keynes. 4 vols. London: Faber and Faber, 1964.

Brykman, Geneviève. "Microscopes and Philosophical Method in Berkeley." In *Berkeley: Critical and Interpretive Essays,* edited by Colin Turbayne, pp. 69–82. Minneapolis: University of Minnesota Press, 1982.

Buetschli, Otto. "Infusoria." In *Klassen und Ordnungen des Thier-Reichs,* edited by Heinrich Georg Bronn, 1:1008-29. Leipzig: Carl Winter, 1880.

Bulloch, William. *The History of Bacteriology.* Oxford: Oxford University Press, 1938. Reprint, New York: Dover, 1979.

Buonanni, Filippo. *Observationes circa viventia . . . cum micrographia curiosa.* Rome: Typis Dominici Antonii Hercules, 1691.

Burnet, Thomas. *Thesaurus medicinae practicae.* London, 1673.

Burtt, Edwin A. *The Metaphysical Foundations of Modern Science.* 2d ed. London: Routledge and Kegan Paul, 1932.

Campanella, Tommaso. *La citta del sole / The City of the Sun.* Translated and edited by Daniel J. Donno. Berkeley and Los Angeles: University of California Press, 1981.

Cardano, Girolamo. *The First Book of Jerome Cardan's De subtilitate.* Translated by Margaret M. Cass. Williamsport, N.Y.: Bayard, 1934.

———. *Opera omnia.* 1663. Reprint, 10 vols., New York: Johnson Reprint, 1967.

Chadarevian, Soraya de. "The Art of Experimenting in Nineteenth-Century German Botany." Manuscript.

Charleton, Walter. *Physiologia Epicuro-Gassendo-Charltoniana.* 1654. Reprint, New York: Johnson Reprint, 1966.

Cherubin d'Orleans, Père C. *La vision parfaite.* Paris: Sebastien Mabre-Cramoisy, 1677.

Clay, Reginald S., and Thomas H. Court. *The History of the Microscope.* London: Griffin, 1932. Reprint, Boston: Longwood, 1979.

Clendening, Logan. *Source Book of Medical History.* New York: P. B. Hoeber, 1942.

Cole, Francis J. "Dr. William Croone on Generation." In *Studies and Essays in the History of Science and Learning,* edited by M. F. Ashley Montagu, pp. 115–35. New York: Schumann, 1944.

———. *Early Theories of Sexual Generation.* Oxford: Clarendon Press, 1930.

———. "Jan Swammerdam." *Nature* 139 (1937): 218–20.

———. "Leeuwenhoek's Zoological Researches." Parts 1 and 2. *Annals of Science* 2 (1937): 1–46, 185–235.

Columella, Lucius Junius Moderatus. *On Agriculture.* Translated by Harrison Boyd Ash. Cambridge, Mass.: Harvard University Press, 1941.

Comenius, John Amos. *The Great Didactic.* Translated and edited by M. W. Keatinge. New York: Russell and Russell, 1967.

———. *Orbis pictus.* 1659. Translated by Charles Hoole, edited by C. W. Bardeen. Reprint, Detroit: Singing Tree Press, 1968.

Copenhaver, Brian P. "Did Science Have a Renaissance?" *Isis* 83 (1992): 387–407.

Cowles, Thomas. "Dr. Henry Power, Disciple of Sir Thomas Browne." *Isis* 20 (1933): 344–66.

Croll, Oswald. *Basilica chymica.* London, 1635.

Croone, William. "De formatione pulli in ovo." *Philosophical Transactions* 7 (1672): 5080.

Culverwell, Nathaniel. *An Elegant and Learned Discourse of the Light of Nature.* 1652. Reprint, New York: Garland, 1978.

Cusanus, Nicholas. *Ueber den Beryll.* Translated from the Latin and edited by Karl Bormann. Hamburg: Felix Meiner, 1977.

Dalenpatius [François de la Plantade]. "Extrait d'un lettre de M. Dalenpatius . . . contenant une decouverte curieuse faite par le moyen du microscope." In

Nouvelles de la république des lettres, p. 552. 1699. Reprint, Geneva: Sklatine, 1966–67.

Daston, Lorraine J. "Baconian Facts, Academic Civility, and the Prehistory of Objectivity." *Annals of Scholarship* 8 (1992): 337–63.

Daston, Lorraine J., and Peter L. Galison. "The Image of Objectivity." *Representations* 40 (1992): 81–128.

Daumas, Maurice. *Scientific Instruments of the Seventeenth and Eighteenth Centuries and Their Makers*. Translated by M. Holbrook. New York: Praeger, 1972. Originally published as *Les instruments scientifiques au XVII et XVIII siècles* (Paris: Presses Universitaires de France, 1953).

Dawson, Virginia P. "The Limits of Observation and the Hypotheses of George Louis Buffon and Charles Bonnet." In *Beyond History of Science: Essays in Honor of R. E. Schofield*, edited by Elizabeth Garber, pp. 107–25. Bethlehem, Pa.: Lehigh University Press, 1990.

Dear, Peter. "Jesuit Mathematical Science and the Reconstitution of Experience in the Early Seventeenth Century." *Studies in the History and Philosophy of Science* 18 (1987): 133–75.

Debus, Allen G., ed. *Science and Education in the Seventeenth Century: The Webster-Ward Debate*. New York: Science History Publications, 1970.

Defoe, Daniel. *A Journal of the Plague Year*. Edited by Louis Landa. London: Oxford University Press, 1969.

de Graaf, Regnier. *De mulierum organis generationi inservientibus*. 1672. Reprint, Nieuwkoop: B. de Graaf, 1965.

Dekker, Thomas. *The Plague Pamphlets of Thomas Dekker*. Edited by Frank P. Wilson. Oxford: Clarendon Press, 1925. Reprint, St. Clair Shores, Wisc.: Scholars' Press, 1971.

Delaporte, François. *Nature's Second Kingdom*. Translated by Arthur Goldhammer. Cambridge, Mass.: MIT Press, 1982. Originally published as *Le second règne de la nature* (Paris: Flammarion, 1979).

Dennis, M. "Graphic Understanding: Instruments and Interpretation in Hooke's *Micrographia*." *Science in Context* 3 (1989): 309–54.

Descartes, René. *Oeuvres*. Edited by Charles Adam and Paul Tannery. 11 vols. Paris: J. Vrin, 1964–74.

———. *Philosophical Writings*. Translated and edited by John Cottingham, Robert Stoothoff, and Dugald Murdoch. 2 vols. Cambridge: Cambridge University Press, 1985.

Dewhurst, Kenneth. *John Locke: Physician and Philosopher*. London: Wellcome Historical Medical Library, 1963. Reprint, New York: Garland, 1984.

Diemerbroeck, Isbrand. *Tractatus de peste*. Amsterdam: J. Blaeu, 1665.

———. *Treatise concerning the Pestilence*. Translated by Thomas Stanton. London, 1722.

Digby, Kenelm. *Two Treatises*. 1645. Reprint, New York: Garland, 1978.

Dijksterhuis, Eduard. *The Mechanization of the World Picture*. Translated by C. Dikshoorn. London: Oxford University Press, 1969.

Disney, Alfred N. *The Origin and Development of the Microscope*. London: Royal Microscopical Society, 1928.

Dobell, Clifford. *Antony van Leeuwenhoek and His "Little Animals."* London: Constable, 1932. Reprint, New York: Dover, 1960.

Dolland, Peter. *Some Account of the discovery made by the late Mr. John Dolland which led to the gradual improvement of refracting telescopes.* London, 1789.

Duchesneau, François. *L'empirisme de Locke.* The Hague: Nijhoff, 1973.

————. "Malpighi, Descartes, and the Epistemological Problems of Iatromechanism." In *Reason, Experiment, and Mysticism in the Scientific Revolution,* edited by Maria Luisa Righini-Bonelli and William R. Shea, pp. 111–30. New York: Science History Publications, 1975.

Eisenstein, Elizabeth. *The Printing Press as an Agent of Change.* 2 vols. Cambridge: Cambridge University Press, 1979.

Espinasse, Margaret d'. *Robert Hooke.* London: Heinemann, 1956.

Feuer, Lewis S. *The Scientific Intellectual.* New York: Basic Books, 1963.

Fludd, Robert. *Tractatus apologeticus.* London, 1617.

Folkes, Martin. "Some Account of Mr. Leeuwenhoek's curious Microscopes, lately presented to the Royal Society. By Martin Folkes, Esq; Vice-President of the Royal Society." *Philosophical Transactions* 32, no. 380 (1722–23): 446–53.

Fontana, Francesco. *Novae coelestium terrestriumque rerum observationes.* Naples, 1646.

Fontenelle, Bernard de. *Conversations on the Plurality of Worlds.* Translated by H. A. Hargreaves. Berkeley and Los Angeles: University of California Press, 1990.

Ford, Brian J. *Single Lens: The Story of the Simple Microscope.* New York: Harper and Row, 1985.

————. *The Leeuwenhoek Legacy.* Forestburgh, New York: Lubrecht and Cramer, 1991.

Foucault, Michel. *The Order of Things.* New York: Pantheon, 1970. Originally published as *Les mots et les choses* (Paris: Gallimard, 1966).

Fouke, Daniel. "Mechanical and 'Organical' Models in Seventeenth-Century Explanations of Biological Reproduction." *Science in Context* 3 (1989): 365–82.

Fournier, Marian. "The Fabric of Life: The Rise and Decline of Seventeenth-Century Microscopy." Ph.D. diss., University of Nijmegen, the Netherlands, 1991.

————. "Huygens' Designs for a Simple Microscope." *Annals of Science* 46 (1989): 575–96.

Fracastoro, Girolamo. "On Contagion." In *Source Book of Medical History,* edited by Logan Clendening. New York: Henry Schurman, 1942.

Freind, John. *The History of Physick.* 4th ed. 2 vols. 1744–50. Reprint, New York: AMS Press, 1975.

Freudenthal, Gad. "Human Felicity and Astronomy: Gersonides' Revolt against Ptolemy" (in Hebrew). *Da'at* 22 (1989): 55–72.

Galen. *Three Treatises on the Nature of Science.* Edited by Michael Frede. Indianapolis: Hackett, 1985.

Galilei, Galileo. *Opere*. 20 vols. Florence: Barbera, 1968.

Garber, Daniel. "Leibniz and the Foundations of Physics: The Middle Years." In *The Natural Philosophy of Leibniz*, edited by Kathleen Okruhlik and James R. Brown, pp. 27–130. Dordrecht: Reidel, 1985.

Gasking, Elizabeth B. *Investigations into Generation, 1651–1828*. Baltimore: Johns Hopkins University Press, 1967.

Gassendi, Pierre. *Opera omnia*. 6 vols. Stuttgart–Bad Canstatt: Fromman, 1964.

Gillispie, Charles Coulton. *The Edge of Objectivity*. Princeton: Princeton University Press, 1960.

———, ed. *Dictionary of Scientific Biography*. 18 vols. New York: Scribner's, 1970.

Ginzburg, Carlo. "High and Low: The Theme of Forbidden Knowledge in the Sixteenth and Seventeenth Centuries." *Past and Present* 73 (1976): 28–41.

Glanvill, Joseph. *Plus ultra*. 1668. Reprint, edited by Jackson I. Cope, Gainesville: Scholars' Facsimiles and Reprints, 1958.

———. *Scepsis scientifica: The Three Versions*. Brighton: Harvester, 1970.

———. *The Vanity of Dogmatizing*. 1661. Reprint, Hildesheim: Olms, 1970.

Goiffon, J. B. *Relations et dissertation sur la peste du Géraudan*. Lyons, 1722.

Govi, Gilberto. "The Compound Microscope Invented by Galileo." *Journal of the Royal Microscopical Society* 9 (1889): 574–98.

Grew, Nehemiah. *The Anatomy of Plants*. London: W. Rawlins, 1682.

———. *The Anatomy of Vegetables Begun*. London: S. Hickman, 1672.

———. *Cosmologia sacra: or, A Discourse of the Universe as it is the Creature and Kingdom of God*. London: Rogers, Smith, and Walford, 1701.

———. *An Idea of a Phytological History Propounded*. London: R. Chiswell, 1673.

Griendel, Johan Franz. *Micrographia nova et curiosa*. Nuremberg, 1687.

Gunther, Robert T. *Early Science in Cambridge*. Oxford: Oxford University Press, 1937. Reprint, London: Dawsons, 1969.

———. *Early Science in Oxford*. 8 vols. Oxford: Clarendon Press, 1921–23. Reprint, London: Dawsons, 1967.

Hacking, Ian. "Do We See through a Microscope?" *Pacific Philosophical Quarterly* 63 (1981): 305–22.

———. *Representing and Intervening*. Cambridge: Cambridge University Press, 1983.

Haller, Albrecht. *Sur la formation du coeur dans le poulet*. Lausanne, 1758.

Harris, John. *Lexicon technicum*. 2 vols. 1710. Reprint, New York: Johnson Reprint, 1966.

———. "Some Microscopical Observations of vast numbers of *Animalcula* seen in Water by John Harris, M.A. Rector of Wincheslea in Sussex, and F.R.S." *Philosophical Transactions* 19, no. 220 (1696): 254–58.

Harting, Pieter. *Das Mikroscop: Theorie, Gebrauch, Geschichte und gegenwaertiger Zustand*. Translated from the Dutch by F. W. Theile. 3 vols. Brunswick, 1859.

Hartsoeker, Nicolaas. *Essai de dioptrique*. Paris: J. Anisson, 1694.

Harvey, William. *Anatomical Exercises*. London, 1653.

————. *Disputations Touching the Generation of Animals.* Translated by G. Whitteridge. Oxford: Blackwell, 1981.

————. *The Works of William Harvey.* Translated by Robert Willis. London: Sydenham Society, 1847.

Harwood, John T. "Rhetoric and Graphics in *Micrographia.*" In *Robert Hooke: New Studies,* edited by Michael Hunter and Simon Schaffer, pp. 119–48. Woodbridge, England: Boydell, 1989.

Hauptmann, August. *Warmer Badt-und Wasser-Schatz.* Leipzig, 1657.

Heilbron, John L. "The Measure of Enlightenment." Introduction to *The Quantifying Spirit in the Eighteenth Century,* edited by Tore Frángsmayr, John L. Heilbron, and Robin E. Rider. Berkeley and Los Angeles: University of California Press, 1990.

————. *Physics at the Royal Society during Newton's Presidency.* Los Angeles: William Andrews Clark Memorial Library, 1983.

Helden, Albert van. *The Invention of the Telescope.* Philadelphia: American Philosophical Society, 1977.

Helmont, Jean Baptiste van. *Oriatrike; or, Physik Refin'd.* Translated by J[ohn] C[handler]. London, 1662.

Henle, Jacob. *A Treatise on General Pathology.* Translated by Henry C. Preston. Philadelphia: Lindsay and Blackiston, 1853.

Heussler, H. *Der Rationalismus des siebzehnten Jahrhundert in seinen Beziehungen zur Entwicklungslehre.* Breslau: Koebner, 1885.

Highmore, Nathaniel. *The History of Generation.* London: Martin, 1651.

Hill, John. *Essays in Natural History and Philosophy.* London, 1752.

————. *Hypochondriasis: A Practical Treatise.* London, 1766.

Hodges, Nathaniel. *Loimologia sive pestis nuperae . . . narratio.* London: Joseph Nevill, 1672.

Hodierna, Giovanni. *L'occhio della mosca discorso fisico.* Palermo, 1644.

Hoefnagel, Jacob. *Archetypa studiaque patris Georgii. Hoefnagelii Jacobus F.* Frankfurt, 1592.

————. *Diversae insectarum volatilium icones ad vivum accuratissime depictae.* Amsterdam, 1630.

Hooke, Robert. *Cometa et microscopium.* London: J. Martyn, 1678.

————. *Micrographia; or, Some Physiological Descriptions of Minute Bodies Made by Magnifying Glasses.* London: J. Martyn and J. Allestry, 1665.

————. *Philosophical Experiments and Observations.* Edited by William Derham. 1726. Reprint, London: Cass, 1967.

————. *Posthumous Works.* Edited by Richard Waller. London: Smith and Walford, 1705.

Houghton, Walter Edwards. "The English Virtuoso." *Journal of the History of Ideas* 2 (1942): 51–73, 190–219.

Hughes, A. F. W. "Studies in the History of Microscopy: The Influence of Achromatism." *Journal of the Royal Microscopical Society* 75 (1955): 1–22.

Hume, David. *An Enquiry concerning Human Understanding.* Edited by Eric Steinberg. Indianapolis: Hackett, 1977.

Hunter, Michael, and Simon Schaffer. *Robert Hooke: New Studies.* Woodbridge, England: Boydell, 1989.

Hutchison, Keith. "Dormitive Virtues, Scholastic Qualities, and the New Philosophies." *History of Science* 29 (1991): 245–78.

———. "What Happened to Occult Qualities in the Scientific Revolution?" *Isis* 73 (1982): 233–53.

Huygens, Christiaan. *Oeuvres complètes.* 22 vols. The Hague: Nijhoff, 1888–1950.

Iliardi, Vincent. "Eyeglasses and Concave Lenses in Fifteenth-Century Florence and Milan: New Documents." *Renaissance Quarterly* 29 (1976): 341–60.

Ivins, William. *Prints and Visual Communication.* Cambridge, Mass.: MIT Press, 1953.

Jacobi, Jolande, trans. and ed. *Paracelsus: Selected Writings.* 2d ed. Princeton: Princeton University Press, 1958.

Joblot, Louis. *Descriptions et usages de plusieurs nouveaux microscopes . . . avec de nouvelles observations. . . .* Paris, 1718.

Joy, Lynn S. *Gassendi the Atomist.* Cambridge: Cambridge University Press, 1989.

Kant, Immanuel. *Gesammelte Schriften.* 29 vols. Berlin: Koeniglich Preussische Akademie der Wissenschaften, 1902–13.

Kepler, Johannes. *The Six-Pointed Snowflake.* Translated by C. Hardie. Oxford: Clarendon Press, 1966.

Kerckring, Theodore. *Spicilegium anatomicum.* Amsterdam, 1670.

Keynes, Geoffrey. *The Life of William Harvey.* Oxford: Clarendon Press, 1978.

King, Edmund. "Several Observations and Experiments on the *Animalcula* in Pepper-water, &c." *Philosophical Transactions* 17 (1693): 861.

King, Henry C. *The History of the Telescope.* London: Charles Griffin, 1955.

Kircher, Athanasius. *Ars magna lucis et umbrae.* Rome, 1646.

———. *Scrutinium physico-medicum contagiosae luis, quae pestis dicitur.* Rome, 1658.

Kirk, G. S., and J. E. Raven. *The Presocratic Philosophers.* Cambridge: Cambridge University Press, 1969.

Klebs, Arnold C., and Eugénie Dorz. *Remèdes contre la peste: Facsimiles, notes et liste bibliographique des incunables sur la peste.* 1925. Reprint, Geneva: Sklatine, 1978.

Koyré, Alexandre. *From the Closed World to the Infinite Universe.* Baltimore: Johns Hopkins University Press, 1953.

———. *Galileo Studies.* Translated by John Mepham. Highlands, N.J.: Humanities Press, 1978. Originally published as *Etudes galileiénnes* (Paris: Hermann, 1966).

———. *Metaphysics and Measurement.* London: Chapman and Hall, 1968.

Kuhn, Thomas S. *The Structure of Scientific Revolutions.* 2d ed. Chicago: University of Chicago Press, 1970.

Lancisi, Giovanni Maria. *Dissertatio historica de bovilla peste.* Rome, 1715.

Lange, Christian. *Opera omnia.* Frankfurt, 1688.

Laudan, Larry. "The Nature and Source of Locke's Views on Hypotheses." *Journal of the History of Ideas* 28 (1967): 211–23.

Le Clerc, Daniel, and Jean-Jacques Manget. *Bibliotheca anatomica.* 3 vols. 2d ed. Geneva, 1699.

Leeuwenhoek, Antoni van. "An abstract of a Letter from Mr. Anthony Leeuwenhoeck of Delft about Generation by an Animalcule of the Male Seed." *Philosophical Transactions* 15, no. 152 (1683): 347–55.

———. "An Abstract of a Letter from Mr. Anthony Leeuwenhoek to Sir CW 22 January 1681/2." *Philosophical Transactions* 13, no. 145 (1682–83): 75.

———. "An Abstract of a Letter of Mr. Leeuwenhoeck Fellow of the R. Society, dated March 30th 1685, to the R.S. Concerning Generation by an Insect." *Philosophical Transactions* 15, no. 174 (1685): 1120–34.

———. *Arcana natura detecta.* Delft: Krooneveld, 1695.

———. *The Collected Letters of Antoni van Leeuwenhoek.* Vols. 1–11. Edited, illustrated, and annotated by a committee of Dutch scientists. Amsterdam: Swets and Zeitlinger, 1939–83.

———. "Concerning the Animalcula in *Semine humano.*" *Philosophical Transactions* 21, no. 255 (1699): 301–8.

———. "Concerning the Texture of Trees, and some remarkable discovery in Wine." *Philosophical Transactions* 11, no. 127 (1676): 653–56.

———. "Monsieur Leewenhoecks Letter to the Publisher, wherein some account is given of the manner of his observing so great a number of little Animals in divers sorts of water, as was deliver'd in the next foregoing Tract: English'd out of the Dutch." *Philosophical Transactions* 12, no. 134 (1677): 844–45.

———. "Observationes D. Anthonii Lewenhoeck, de Natis è semine genitali Animalculis." *Philosophical Transactions* 12, no. 142 (1677–78): 1040–43.

———. "Observations, communicated to the Publisher by Mr. Antony van leeuwenhoeck, in a Dutch Letter of the 9th of Octob. 1676, here English'd: Concerning little Animals by him observed in Rain-Well-Sea- and Snow water; as also in water wherin Pepper had lain infused." *Philosophical Transactions* 12, no. 133 (1677): 821–31.

———. *Opera omnia seu arcana naturae.* 4 vols. Leyden: Langerak, 1722.

———. "Other Microscopical observations, made by the same, about the texture of the Blood, the Sap of some Plants, the Figure of Sugar and Salt, and the probable cause of the difference of their Tasts." *Philosophical Transactions* 10, no. 117 (1675): 380–85.

Legrand, Christian. "De l'invention des lunettes à la 'Dioptrique': Pratique et exposition de la méthode chez Descartes." 1991. Manuscript.

Leibniz, Gottfried Wilhelm. *Discourse on Metaphysics.* Translated by George R. Montgomery. LaSalle, Ill.: Open Court, 1973.

———. *Leibniz: Philosophical Papers and Letters.* Edited by Leroy Loemker. 2d ed. Dordrecht: Reidel, 1969.

———. *Philosophische Schriften.* Edited by Carl Immanuel Gerhardt. 7 vols. 1875. Reprint, Hildesheim: Olms, 1960–61.

———. *Sämtliche Schriften und Briefe.* Berlin: Deutsche Akademie der Wissenschaften, 1923–.

Lemnius, Levinus. *The Secret Miracles of Nature.* London, 1658.

Lesser, Friedrich Christian. *Insecto-theologia.* Frankfurt and Leipzig, 1738.

Lindberg, David C., and Nicholas H. Steneck. "The Sense of Vision and the Origins of Modern Science." In *Science, Medicine, and Society in the Renais-*

sance: Essays to Honor Walter Pagel, edited by Allen G. Debus, pp. 29–45. London: Science History Publications, 1972.

Lindquist, Svante. "A Wagnerian Theme in the History of Science: Scientific Glassblowing and the Role of Instrumentation." In *Solomon's House Revisited*, edited by Tore Frängsmayr, pp. 160–83. Canton, Mass.: Science History Publications, 1990.

Lloyd, C. F. "Shadwell and the Virtuosi." *Proceedings of the Modern Language Association* 44 (1929): 472–94.

Locke, John. "Anatomie." Translated by Kenneth Dewhurst. *Medical History* 2 (1958): 3–8.

———. *The Educational Writings of John Locke.* Edited by James L. Axtell. Cambridge: Cambridge University Press, 1968.

———. *Essay concerning Human Understanding.* Edited by Peter H. Nidditch. Oxford: Clarendon Press, 1975.

Loeb, Louis. *From Descartes to Hume.* Ithaca: Cornell University Press, 1981.

Macasius, J. C. *Specimen pathologiae animatae de morbillis.* Leipzig, 1660.

M.A.C.D. *Suite du système d'un medecin anglois, sur la guerison des maladies.* Paris, 1727.

———. *Système d'un medecin anglois sur la cause de toutes les especes des maladies.* Paris, 1726.

Malebranche, Nicolas. *Dialogues on Metaphysics.* Translated by Willis Doney. New York: Abaris, 1980.

———. *Oeuvres complètes.* Edited by André Robinet. 20 vols. Paris: J. Vrin, 1958.

———. *The Search after Truth.* Translated by Thomas M. Lennon and Paul Olscamp. Columbus: Ohio State University Press, 1980.

Malpighi, Marcello. *Anatomes plantarum.* 2 vols. London: J. Martyn, 1675–79.

———. *Dissertatio epistolica de formatione pulli in ovo.* London: J. Martyn, 1673.

———. *Opera omnia.* London: Scott and Wells, 1686.

Marten, Benjamin. *A New Theory of Consumptions.* London, 1720.

Martin, T. M. H. "Sur des instruments d'optique faussement attribués aux anciens." *Bulletino di bibliografia e di storia della scienze matematiche e fisiche* 4 (1871).

Marx, Jacques. "L'art d'observer au XVIII siècle: Jean Sénebier et Charles Bonnet." *Janus* 61 (1974): 201–20.

Massuet, P. *De generatione ex animalculo in ovo.* Leyden: Verbeek, 1729.

Matthen, Mohan. "Empiricism and Ontology in Ancient Medicine." In *Method, Medicine, and Metaphysics: Studies in the Philosophy of Ancient Science*, edited by R. James Hankinson, pp. 99–122. Edmonton: Academic Printing and Publishing, 1988.

Mayall, John. "Cantor Lectures on the Microscope." *Journal of the Royal Society of Arts* 34 (1887): 987–97, 1007–21; 36 (1888): 1149–59, 1164–72.

McCormick, J. B., and Gerald L'E. Turner. *The Atlas Catalog of Replica Rara, Ltd., Antique Microscopes.* Chicago: Replica Rara, 1975.

Mead, Richard. *A Short Discourse concerning Pestilential Contagion*. London, 1720.

Meinel, Christoph. "Das letzte Blatt im Buch der Natur." *Studia Leibnitiana* 20 (1988): 1–18.

Merchant, Carolyn. *The Death of Nature*. New York: Harper and Row, 1980.

Mersenne, Marin. *La veritez des sciences*. 1625. Reprint, Stuttgart–Bad Canstatt: Fromann, 1969.

Miall, Louis C. *The Early Naturalists: Their Lives and Work*. London: Macmillan, 1912.

Michelet, Jules. "L'Insecte." In *Oeuvres complètes* 17:261–453. Paris: Flammarion, 1971.

Middleton, William Edgar Knowles. *The Experimenters: A Study of the Accademia del Cimento*. Baltimore: Johns Hopkins University Press, 1971.

Miscellanaea curiosa medico-physica academiae naturae curiosum. Vol. 2. Frankfurt and Leipzig, 1678–79.

Moffett, Thomas. *Theater of Insects*. Translated and edited by Theodore de Mayherne. Vol. 4 of *The History of Four-Footed Beasts and Serpents*, by Edward Topsell. London, 1658.

Monconys, Balthasar de. *Journal des voyages de Monsieur de Monconys*. 3 vols. in 2. Lyons: Boissat and Remeus, 1665–66.

Needham, John Turberville. *A History of Embryology*. Cambridge: Cambridge University Press, 1934.

Nelson, Edward M. "A Bibliography of Works (Dated Not Later than 1700) Dealing with the Microscope and Other Optical Subjects." *Journal of the Royal Microscopical Society* 22 (1902): 20–23.

Newton, Isaac. *Opticks*. 4th ed. 1730. Reprint, New York: Dover, 1952.

Nicolson, Marjorie. *Science and Imagination*. Ithaca: Cornell University Press, 1956.

Pagel, Walter. *Aspects of van Helmont's Science and Medicine*. Baltimore: Johns Hopkins University Press, 1944.

———. *J. B. van Helmont, Reformer of Science and Medicine*. Cambridge: Cambridge University Press, 1982.

———. *New Light on William Harvey*. Basel: S. Karger, 1976.

———. *Paracelsus*. 2d ed. Basel: S. Karger, 1982.

———. "Religious Motives in the Medical Biology of the Seventeenth Century." *Bulletin of the Institute of the History of Medicine* 3 (1935): 97ff.

Paracelsus [Philippus Aureolus Theophrastus Bombast von Hohenheim]. *Sämtliche Werke*. Edited by B. Ascher. Jena: Fischer, 1932.

Park, Katherine, and Lorraine J. Daston. "Unnatural Conceptions: The Study of Monsters in Sixteenth- and Seventeenth-Century France and England." *Past and Present* 92 (1981): 21–54.

Pascal, Blaise. *Oeuvres complètes*. Edited by Louis Lafuma. Paris: Editions du Seuil, 1963.

Paullini, Christian Franz. *Cynographia curiosa*. Nuremberg, 1685.

———. *Disquisitio curiosa an mors naturalis plerumque sit substantia verminosa*. Frankfurt and Leipzig, 1703.

Philipp, Wolfgang. *Das Werden der Aufklärung in theologiegeschichtlicher Hinsicht.* Göttingen: Vandenhoech and Ruprecht, 1957.

Porta, Giambattista della. *Natural Magick.* London: Young and Speed, 1658.

Power, Henry. *Experimental Philosophy, in Three Books, Containing New Experiments, Microscopical, Mercurial, Magnetical.* 1664. Reprint, New York: Johnson Reprint, 1966.

Price, Derek J. de Solla. "Of Sealing Wax and String." *Natural History* 93 (1984): 49–56.

————. "Philosophical Mechanism and Mechanical Philosophy: Some Notes towards a Philosophy of Scientific Instruments." *Annali dell'Instituto e Museo di storia della scienza* 5 (1980): 75–85.

Puget, L. de. *Observations sur la structure des yeux de divers insectes.* Lyons, 1706.

Punnett, Reginald Crundall. "Ovists and Animalculists." *American Naturalist* 62 (1928): 481ff.

Quarles, Francis. *Emblems, Divine and Moral.* London: Tegg, 1859.

Rádl, Emmanuel. *Geschichte der biologischen Theorien in der Neuzeit.* 2 vols. 2d ed. Hildesheim: Olms, 1970.

Recorde, Robert. *The Pathway to Knowledge.* London, 1551.

Redi, Francesco. *Experiments on the Generation of Insects.* 1668. Translated by M. Bigelow. 1909. Reprint, New York: Kraus Reprint, 1969.

Regis, Pierre Sylvain. *Cours entier de philosophie.* 1691. Reprint, New York: Johnson Reprint, 1970.

Review of "An Account of *Cochlearia Curiosa, or the Curiosities of Scurveygrass,* written in Latin by Dr. Andr. Molimbrochius of Leipsig, and English'd by Dr. Th. Sherley Physician in ordinary to his Majesty." *Philosophical Transactions* 11, no. 125 (1676): 621.

Richards, Owain W. "The History of the Microscope: Selected References." *Transactions of the American Microscopical Association* 68 (1949): 55.

Righini-Bonelli, Maria Luisa, and William R. Shea. *Reason, Experiment, and Mysticism in the Scientific Revolution.* New York: Science History Publications, 1975.

Roe, Shirley. *Matter, Life, and Generation.* Cambridge: Cambridge University Press, 1981.

Roger, Jacques. *Les sciences de la vie dans la pensée française du XVIIIème siècle.* 2d ed. Paris: A. Collin, 1971.

Romanell, Patrick. *John Locke and Medicine.* Buffalo, N.Y.: Prometheus, 1984.

Ronchi, Vasco. "The General Influence of the Development of Optics in the Seventeenth Century on Science and Technology." In *Vistas in Astronomy,* edited by Arthur Beer, 9:123–33. Oxford: Pergamon, 1968.

————. "The Influence of the Early Development of Optics on Science and Philosophy." In *Galileo: Man of Science,* edited by Ernan McMullin, pp. 195–206. New York: Basic Books, 1967. Reprint, Princeton Junction: Scholars' Bookshelf, 1988.

Roseboom, Maria. *Microscopium.* Leyden: Rijksmuseum voor de Geschiedenis der Natuurwetenschappen, 1956.

Rösel von Rosenhof, A. J. *Die monatlich-herausgegebenen Insecten-Belustigungen*. 4 vols. Nuremberg, 1746–61.

Rosen, Edward. "The Invention of Eyeglasses." *Journal of the History of Medicine and Allied Sciences* 11 (1956): 13–46, 183–218.

Ruestow, Edward G. "Images and Ideas: Leeuwenhoek's Perception of the Spermatozoa." *Journal of the History of Biology* 16 (1985): 185–224.

———. "Leeuwenhoek and the Campaign against Spontaneous Generation." *Journal of the History of Biology* 17 (1984): 225–48.

———. "Piety and the Defense of the Natural Order: Swammerdam on Generation." In *Religion, Science, and Worldview*, edited by Margaret J. Osler and Paul Farber, pp. 217–41. Cambridge: Cambridge University Press, 1985.

Sarton, George. *The Appreciation of Ancient and Medieval Science during the Renaissance (1450–1600)*. New York: A. S. Barnes, 1955.

Schmitz, E. H. *Handbuch zur Geschichte der Optik*. 4 vols. Bonn: Wayenborgh, 1981.

Schott, Gaspar. *Magia universalis*. Würzburg, 1657–59.

———. *Technica curiosa*. Würzberg, 1664.

Schrader, Friedrich. *De microscopiorum usu in naturali scientia et anatome*. Göttingen, 1681.

Shadwell, Thomas. *The Virtuoso*. Edited by Marjorie Nicolson. London: Arnold, 1966.

Shapin, Steven, and Simon Schaffer. *Leviathan and the Air-Pump: Hobbes, Boyle, and the Experimental Life*. Princeton: Princeton University Press, 1985.

Silverstein, Arthur M. *A History of Immunology*. San Diego: Academic Press, 1989.

Simpson, A. D. C. "Robert Hooke and Practical Optics." In *Robert Hooke: New Studies*, edited by Michael Hunter and Simon Schaffer, pp. 33–62. Woodbridge, England: Boydell, 1989.

Singer, Charles. "Notes on the Early History of Microscopy." *Proceedings of the Royal Society of Medicine* 7 (1913–14): 247–79.

———. *A Short History of Scientific Ideas to 1900*. London: Oxford University Press, 1970.

———. "Steps Leading to the Invention of the First Optical Apparatus." In *Studies in the History and Method of Science*, edited by Charles Singer, 2:385. London: W. Dawson, 1955. Reprint, New York: Arno, 1975.

Singer, Charles, and Dorothea Singer. "The Development of the Doctrine of *Contagium Vivum*, 1500–1750." In *Proceedings of the XVIIth International Medical Conference*, pp. 187–206. London, 1914.

Sloan, Philip. "Organic Molecules Revisited." In *Buffon 88*, edited by Jean Gayon et al. pp. 415–38. Paris: Vrin, 1992.

"Some Considerations of an Observing person in the Country upon Numb. 133 of these Tracts, sent in a Letter to the Publisher of May 2, 1677." *Philosophical Transactions* 12, no. 136 (1677): 890–92.

Spinoza, Benedict. *Ethics*. Vol. 1 of *Collected Works*. Translated by Edmund Curley. Princeton: Princeton University Press, 1985.

Sprat, Thomas. *The History of the Royal Society.* Edited by Jackson I. Cope and Harold W. Jones. St. Louis: Washington University Press, 1958.

Stelluti, Francesco. *Persio tradutto in verso sciolto e dichiarato.* Rome, 1630.

Stimson, Dorothy. *Scientists and Amateurs.* New York: Henry Schurman, 1948.

Swammerdam, Jan. *The Book of Nature.* Translated by Thomas Flloyd. London, 1758.

Temkin, Oswei. "The Elusiveness of Paracelsus." *Bulletin of the History of Medicine* 26 (1952): 201–17.

Thomas, Keith. *Religion and the Decline of Magic.* New York: Scribner's, 1971.

Thorndyke, Lynn. *A History of Magic and Experimental Science.* 8 vols. New York: Macmillan, 1923–58.

Topsell, Edward. *The History of Four-Footed Beasts and Serpents.* London, 1658.

Torrey, H. B. "Athanasius Kircher and the Progress of Medicine." *Osiris* 5 (1938): 246–75.

Turner, Gerald L'E. "The Microscope as a Technical Frontier in Science." In *Historical Aspects of Microscopy,* by Savile Bradbury and Gerald L'E. Turner, pp. 173–99. Cambridge: Heffer and Sons, 1967.

———. "Microscopical Communication." *Journal of Microscopy* 100 (1974): 3–20.

Vallisnieri, Antonio. *Opere fisico-medichie.* Edited by Antonio Vallisnieri (the younger). 3 vols. Venice, 1733.

Varro, Marcus Terentius. *On Agriculture.* Translated by William Davis Hooper. Cambridge, Mass.: Harvard University Press, 1934.

Vickers, Brian. "Analogy vs. Identity: The Rejection of Occult Symbolism, 1580–1680." In *Occult and Scientific Mentalities in the Renaissance,* edited by Brian Vickers, pp. 95–163. Cambridge: Cambridge University Press, 1984.

Vierblaetterichter Wunder-Klee. Nuremberg: C. Gerhard, 1667.

Waard, Cornelis de. *De Uitvinding des Verrekijkers.* The Hague: Smits, 1906.

Walker, Daniel P. *Spiritual and Demonic Magic from Ficino to Campanella.* 2d ed. Notre Dame: University of Notre Dame Press, 1975.

Ward, Seth, and John Wilkins. *Vindiciae academiarum.* In *Science and Education in the Seventeenth Century: The Webster-Ward Debate,* edited by Allen G. Debus. New York: Science History Publications, 1970.

Wear, Andrew. "William Harvey and the Way of the Anatomists." *History of Science* 21 (1983): 223–48.

Webster, Charles. *The Great Instauration: Science, Medicine, and Reform, 1626–1660.* London: Duckworth, 1975.

Webster, John. *Academiarum examen.* In *Science and Education in the Seventeenth Century: The Webster-Ward Debate,* edited by Allen G. Debus. New York: Science History Publications, 1970.

Westfall, Richard S. *The Construction of Modern Science.* Cambridge: Cambridge University Press, 1971.

Wilkins, John. *An Essay towards a Real Character and a Philosophical Language.* 1668. Reprint, Menston, England: Scolar, 1968.

Willis, Thomas. *Diatribe duae medico-philosophicae.* London: Martin, Allestry, and Dicas, 1662.

Wilson, Catherine. "Berkeley and the Microworld." *Archiv fuer Geschichte der Philosophie* 76 (1994): 37–64.

———. "The Microscope and the Occult." *Journal of the History of Ideas* 49 (1988): 85–108.

Wilson, Frank P. *The Plague in Shakespeare's London.* Oxford: Oxford University Press, 1981.

Wilson, Margaret. "Berkeley and the Essences of the Corpuscularians." In *Essays on Berkeley,* edited by John Foster and Howard Robinson, pp. 131–48. Oxford: Clarendon Press, 1985.

Wither, George. *The History of the Pestilence.* 1625. Reprint, edited by J. M. French, Cambridge, Mass.: Harvard University Press, 1932.

Wolff, Caspar Friedrich. *Theoria generationis.* Halle, 1759.

Wright, Edward. "Microscopical Observations: In a Letter from Edward wright, Esq; to Mr. Peter Collinson, F.R.S. dated at Paris, Decemb. 26, 1755." *Philosophical Transactions* 49, part 2 (1756): 553–58.

Yolton, John. *Locke and the Compass of Human Understanding.* Cambridge: Cambridge University Press, 1970.

———, ed. *Philosophy, Religion, and Science in the Seventeenth and Eighteenth Centuries.* Rochester, N.Y.: University of Rochester Press, 1990.

Yost, R. M. "Locke's Rejection of Hypotheses about Sub-microscopic Events." *Journal of the History of Ideas* 12 (1951): 111–30.

Zahn, Johann. *Oculus artificialis teledioptricus, sive telescopium.* Würzburg, 1685–86.

Zanobio, Bruno. "L'immagine filamentoso-reticolare nell'anatomia microscopica dal 17 al 19 secolo." *Physis* 2 (1960): 299–317.

———. "Micrographie illusoire et theories de la structure de la matiere vivante." *Clio medica* 6 (1971): 25–40.

Index